The Collector's Book of Fluorescent Minerals

The Collector's Book of Fluorescent Minerals

Manuel Robbins

VNR VAN NOSTRAND REINHOLD COMPANY
NEW YORK CINCINNATI TORONTO LONDON MELBOURNE

Library of Congress Catalog Card Number: 82-17346
ISBN: 0-442-27506-4

Manufactured in the United States of America

Published by Van Nostrand Reinhold Company Inc.
135 West 50th Street, New York, N.Y. 10020

Van Nostrand Reinhold
480 Latrobe Street
Melbourne, Victoria 3000, Australia

Van Nostrand Reinhold Company Limited
Molly Millars Lane
Wokingham, Berkshire, England

15 14 13 12 11 10 9 8 7 6 5 4 3 2 1

Library of Congress Cataloging in Publication Data

Robbins, Manuel.
The collector's book of fluorescent minerals.

Includes index.
1. Mineralogy--Collectors and collecting. 2. Fluorescence. I. Title.
QE364.2.F5R6 1983 549'.125 82-17346
ISBN 0-442-27506-4

This book is dedicated to my wife
Renee Robbins

Preface

Over the last several decades, the number of people who are actively involved in the hobby or science of mineral collecting has grown at an increasing pace. In response to the growing demand for information which this large and active group has created, a number of books have been published dealing with mineralogy. As a result, the reader now has a choice among mineral locality guides, field handbooks, photo collections, or books dedicated to the systematic description of minerals.

However, as interest in mineralogy has grown, as collectors have become increasingly knowledgeable and aware of mineralogy in its many facets, the need for more specialized information has also grown. Nowhere is this need greater than in the subject of the fluorescence of minerals. The number of collectors who now maintain a fluorescent collection is substantial, interest is constantly increasing, and manufacturers have recently responded by the introduction of new ultraviolet equipment with major improvements in utility and performance. Yet when the collector searches for any information on this subject, little will be found. He or she will seek in vain for the answers to questions which present themselves as interest in fluorescent minerals grows and matures. Which minerals fluoresce? Where are fluorescent minerals found? What makes a mineral fluoresce? Why does ultraviolet light produce fluorescence? What is an activator, and how does it contribute to fluorescence? On these matters, the available mineralogy books are largely silent.

These same questions inevitably arose as my own interest in fluorescent minerals developed. I have been a collector of Franklin minerals and, of course, I gathered as many of the fabulous fluorescent minerals of Franklin as could be obtained. Then, slowly at first, I began to recognize that fine fluorescent minerals could be found elsewhere and the scope of my collecting broadened. This required that I search for fluorescent minerals at mineral shows, in the shops of mineral dealers, and in the field, looking for fluorescent surprises from far off places. Time after time, I found them. Most of the fluorescent minerals found this way were surprises if only because so little has been written about fluorescent minerals that each find was a discovery, a revelation. As my collection grew, so did the notes which I kept. The notes described not only fluorescent minerals collected or purchased, but also those seen, those written about, or those described to me by fellow collectors. These provided a target list for future acquisition.

Soon enough, the question of the cause of fluorescence in minerals became important. Here a different type of search was required, a search through the technical books and journals which report the investigations of physicists and physical chemists into the causes of luminescence, which is simply fluorescence and phosphorescence in our terms. Discussions with scientists working in this field were also helpful and provided useful information.

Until that time, I had been motivated by my own interest and curiosity, but as time went on, as I talked with many other mineral collectors, it slowly became clear that there existed a widespread need for information on mineral fluorescence. To fill this void, a book would be required, and so, this book was written.

The book is addressed primarily to the mineral collector. The beginning collector should find here a guide and help to collection building. He or she will find information on the type of ultraviolet light to use, the minerals to look for, where to find them, and the way to organize the collection. It is likely that the advanced collector also may find here useful or interesting information beyond his or her present knowledge. If the beginning collector is aided, if the more advanced collector finds here the answers to some things he or she has been wondering about, I feel that this book will have fulfilled its purpose.

I have visited and collected minerals at a number of localities mentioned in this book. These visits have been particularly useful in preparing Chapter 5. In preparing that chapter, I also called upon a number of people who have provided help, especially by patiently answering a number of questions or reviewing material relating to the locality or mineral formations in which they specialize. I appreciate their help, and I want to thank Fred Totten of the Governeur Talc Company at Balmat, New York, and C. MacDonald Grout and Bill Lorraine, geologists of the St. Joe Resources Company, whose offices are also at Balmat, New York. Similarly, thanks are extended to John Baum, retired geologist of the New Jersey Zinc Company at Franklin, and Jim Minette of the U.S. Borax operation at Boron, California. Both Fred Divoto and Ruth Kirkby, dedicated collectors of Crestmore minerals, have also been helpful with regard to the geology and mineralogy of that remarkable location. At Searles Lake, Gail Moulton and Steve Mulqueen, geologists of the Kerr-McGee Company, extended every courtesy, and their help is acknowledged. Richard Gaines of Pottstown, Pennsylvania, also provided help for which I am grateful.

A number of these people also allowed me to examine under the ultraviolet light the office or reference collections which are maintained under their supervision and to make useful notes on the fluorescence of some minerals which I might not have had a chance to see otherwise. Others helped in this way. Ewald Gerstmann of Franklin, New Jersey, permitted me to examine a number of specimens in the Gerstmann-SPEX collection of Franklin minerals. Dick Bostwick and Warren Miller allowed me to inspect their personal Franklin fluorescent collections. Tom Peters allowed me to examine the extensive collection of zeolite and other minerals in the Paterson Museum at Paterson, New Jersey. The courtesy and help of all of these people are acknowledged with thanks.

David Williams of the Center for Visual Science at the University of Rochester reviewed a number of questions relating to Chapter 9. His help is appreciated. Maria Crawford of Bryn Mawr College devoted substantial time on many occasions in order to identify a number of silicate minerals described in this book. I also want to thank Neil Yocum of the RCA Research Laboratories at Princeton, New Jersey. His years of experience in developing both phosphors and laser materials have given him deep insight into the nature of fluorescence and the activators involved. He has helped me on many occasions when some point in the technical literature was particularly unclear.

I also want to thank Tom Warren of Ultra-violet Products for recounting his experiences in Chapter 4. For that chapter, Dick Bostwick was kind enough to provide an account of his experiences in the Sterling mine. Don Newsome, first president of the Fluorescent Mineral Society, provided the graphs of the spectra of a number of fluorescent minerals which appear in Chapter 9. I extend thanks to them all.

Color photographs of a number of fluorescent minerals are provided as a means of illustrating the beauty and variety to be found among fluorescent minerals, and as an aid to identification. Most of these photographs are of specimens in the author's collection. These were photographed at f/4.5 on Kodachrome ASA 64 with time exposures ranging between 7 and 50 seconds. The minerals were illuminated by hand held ultraviolet lights at several inches distance. Other photographs are of specimens in other collections, and their owners were kind enough to furnish these photographs for use here. Tom Warren of Ultra-Violet Products furnished color pictures of specimens 27, 29, 50, 53, 55, and 56. Specimen 61 is in the Ray Vajdik collection. Specimens 39 and 60 are in the Charles Weed collection. Specimens 35, 63, and 64 are in the Dick Bostwick collection, and specimen 62 is in the Warren Miller collection. Thanks are extended to all for the use of these photographs.

Contents

1
Background

Man is one of those few creatures who can see the world in color. Almost as far as we are able to trace the presence of man in history and prehistory, we find evidence that he has employed color to adorn his person and his surroundings. Tens of thousands of years ago, Cro-Magnon created cave paintings using the colors of minerals as pigments, and before him, the Neanderthal painted the bones of the dead with red mineral pigment. In the ancient world, colored stones and minerals were used in jewelry, in the decoration of buildings, and as coloring agents in glasses and enamels.

In gems, color was a dominant consideration. In ancient times, ruby or sapphire or emerald meant any hard and gemmy stone of red, blue, or green color. This passion for color in mineral form undoubtedly reached its highest state among the pre-Colombian Indian peoples of Middle America, particularly the Mayan and Aztec, for whom green jade was the most precious of materials. Numerous Mayan burials have been found containing a treasure trove of bright green jadeite necklaces, ear spools, bracelets, and other adornments.

Today, on a more modest scale, every mineral collector probably favors, in particular, those specimens in his collection which are vivid in color, whether crystals of azurite, rhodonite, crocoite, or crystal specimens of the gem minerals.

It is thus interesting to realize that over the tens of thousands of years that man has sought color in minerals, the most vivid, brilliant and dramatic – and at the same time the most subtle and varied – color phenomenon in minerals was unknown until recent times. This is the phenomenon of fluorescence.

The recognition of fluorescence in minerals, the formation of fluorescent collections, and the rapid growth of this aspect of the ancient interest in minerals have been substantially recent occurrences, since they are a product of the electrical and, more recently, the electronic age. The development of electricity was required in order to provide the various practical sources of ultraviolet light which are required to produce fluorescence; more recently, electronic circuitry has made portable field lights practical.

Today, it is probably true that the majority of the millions of rock and mineral collectors own an ultraviolet light with which they can investigate the fluorescent response of minerals in their collection. Many collectors maintain a display collection of fluorescents, and some collectors specialize in this branch of the mineral hobby. The reason they do so is directly evident once the ultraviolet light is directed at a fluorescent mineral. Particularly when short wave ultraviolet is used, brilliant and intense colors may spring forth from a mineral specimen which may be unattractive and unpromising in ordinary light, or a more subdued and pastel effect may emerge under long wave light. When a collection includes a selected variety of specimens arranged so that color, intensity, and hue are varied yet properly balanced, the effect is astonishing. Indeed, the fluorescent collection is as much a field of art and aesthetics as it is a field for scientific pursuit.

What is fluorescence and what is its cause? Fluorescence is the visible light produced by certain minerals when these minerals are illuminated by the invisible light rays of an ultraviolet or black light source. Whatever the color, the result is primarily a property of those minerals whose makeup allows them to absorb invisible ultraviolet and to reemit some of this energy as visible light. Impurities known as "activators" play an important role in this process in many minerals. Phosphorescence, which is the continued glow of light after the source of ultraviolet is turned off or removed, is very closely related to fluorescence. Probably all minerals which fluoresce also phosphoresce, though it is only in some that the intensity and duration of this afterglow are sufficient to be easily seen.

Fluorescence-like effects in minerals can also be produced by means other than the use of ultraviolet light and these should be briefly mentioned. Streams of electrons directed at some minerals will produce an output of visible light. This effect is known as cathodoluminescence and is exploited to produce the visible picture in the TV picture tube. Radium and X-rays will produce a similar result called radioluminescense. This is exploited to render X-ray pictures visible. Cathodoluminescence and radioluminescence are only of passing interest to most mineral collectors since the apparatus needed to produce the excitation is elaborate and generally unavailable to the collector.

In some minerals, a momentary burst of light can be produced by heating the mineral to a sufficiently high temperature. This effect is known as thermoluminescence. It has been used by geochemists as a means of dating the time of formation of some minerals since the light produced is the result of energy stored in the mineral, derived from the environment of the mineral at a known rate over geological time. Thermoluminescence is also used by archaeologists to date pottery fragments.

Finally, some minerals will produce light when scratched or struck. This is known as triboluminescence and is familiar to any collector who, breaking quartz, feldspar, or certain other rocks in the dark, notes the flash of light at the point of impact of the hammer.

Some of these luminous effects in minerals were known in ancient times, while others are comparatively modern discoveries. Triboluminescence has probably been recognized for as long as men have worked ores in the deep dark of the mines – at least several thousand years. Phosphorescence was probably noticed in early times by lime makers or dabblers in alchemy who chanced to take certain calcite, or a baked limestone or barite, from sunlight into the dark. It has also been known for centuries that sunlight could produce phosphorescence in some diamonds. The ultraviolet in sunlight is the stimulus for such phosphorescence. Cathodoluminescence and radioluminescence were discovered in the nineteenth century in the course of experiments with electricty and radioactivity.

Fluorescence produced by ultraviolet light is also a discovery of the nineteenth century. About 1800, various experimenters working with Newton's prism discovered that the region just past the visible violet of the spectrum contained radiation since it would darken silver salts, which acted as a primitive photographic plate. Thus was ultraviolet, meaning the region beyond the violet, discovered. The German poet Goethe now enters the scene. Goethe (1749–1832) is certainly much more widely known as a writer and poet than as a scientist, but in fact, he was passionately devoted to the study of the natural sciences, and we may recall that the mineral goethite was named in his honor. He was perhaps the first to recognize and record the fact that ultraviolet light can produce fluorescence in minerals. In his writings on the subject in 1810, which culminated perhaps two decades of study of what we would now call fluorescence, he stated that "beyond the violet where scarcely any color can be seen, the phosphor gave a vivid brilliance . . . ." About the same time, another researcher discovered that the radiation from the electric spark, now known to be a rich source of ultraviolet, would produce fluorescence.

We now move to the remarkable Becquerel family of France, three generations of scientists whose work in fluorescence covers a good part of the nineteenth century. Antoine Becquerel (1788–1878) and his son Edmond investigated the response of numerous fluorescent materials to various wavelengths of light, and measured the spectrum of the emitted fluorescent light. Edmond Becquerel (1820–1891) also discovered the red fluorescence of calcite and determined that this fluorescence was due to the presence of some 2.7% manganese in the calcite. This was perhaps the earliest certain discovery of an activator in a fluorescent mineral. He also invented the phosphoroscope, a device for measuring the duration of phosphorescence. In this apparatus, phosphorescence as short as a ten thousandth of a second could be measured.

Then, late in the nineteenth century, Henri Becquerel (1852–1908), son of Edmond, made one of the most important discoveries in the history of science. He was investigating whether a fluorescent material might give off some form of invisible radiation in addition to visible light. For this purpose, he exposed various materials to sunlight and then placed them on a photographic plate to study the response. At one point, he happened to try a material containing uranium. While waiting for a sunny day, he placed the uranium on top of a still covered photographic plate and put it away in a drawer. Some time later, forgetting that the plate had not been exposed, he developed it. To his surprise, the plate was darkened beneath the place where the uranium had been. This response had nothing to do with fluorescence or phosphorescence. It was due solely to invisible penetrating radiation emitted by the uranium. Thus, from what was intended as an experiment in fluorescence, radioactivity was discovered and the modern world of atomic physics was born. The mineral becquerelite was named in his honor, a fitting tribute to the man whose experiments in fluorescence lead to the discovery of radioactivity. Becquerelite is both fluorescent and radioactive.

It is worth mentioning at this point that there is no connection between fluorescence and radioactivity. It is purely coincidental that many uranium minerals are both fluorescent and mildly radioactive. Other fluorescent minerals are not radioactive and present no hazard.

We now return to the middle of the nineteenth century and George Stokes at Cambridge, England. Stokes (1820–1903) had been investigating the fluorescence of various materials. He recognized, as had others, that the light given off was not simply a reflection or an internal scattering of the light directed at the mineral. Casting about for a name for this behavior, he gave it our present day term "fluorescence" after the mineral fluorite, many specimens of which showed this response so prominently. He did this by analogy with the term "opalescence" which is named after opal.

Stokes also made the important observation that the light given off in fluorescence is always of longer wavelength than the light which, directed at the material, produces the fluorescence. Thus, a blue light might produce a red fluorescence, but a red light cannot produce a blue fluorescence. While certain special exceptions are known, this rule holds so widely that it has been given the name Stokes law of fluorescence. This law plays a key role in the modern explanation of fluorescence, which is based on the quantum theory of physics as developed in the early part of the twentieth century.

By the twentieth century, the investigation of fluorescence and phosphorescence by mineralogists and mineral collectors was under way. Early in the twentieth century, Charles Baskerville and George Kunz, after whom the mineral variety kunzite was named, began a massive investigation of the fluorescent response of the mineral collection at the American Museum of Natural History, thereby creating one of the first catalogues of fluorescent minerals. In this research, the ultraviolet fluorescence of willemite, colemanite, hanksite, glauberite, hydrozincite, hyalite, fluorite, gypsum, topaz, pectolite, wernerite, wollastonite, calcite, witherite, strontianite, aragonite, cerussite, and other minerals as well, was noted. Mineralogists in Europe soon added the results of their own surveys. By the 1920s, spark lights and other sources of ultraviolet were becoming available to the mineral collector. The building of fluorescent mineral collections was under way.

2
What Is Ultraviolet?

THE ULTRAVIOLET SPECTRUM

Scheelite fluoresces bright blue only under short wave ultraviolet light, and ruby fluoresces bright red primarily under long wave ultraviolet. Some minerals will fluoresce one color under short wave and a different color under long wave. Calcite from Terlingua, Texas, is a spectacular example of this, fluorescing bright blue under short wave, bright pink under long wave. Other fluorescent minerals are more or less impartial. Some willemites and some uranium minerals will fluoresce green; many fluorites will fluoresce blue; many calcites will fluoresce red under either short or long wave.

What is this ultraviolet that produces these spectacular responses in minerals, and what is the distinction between short wave and long wave ultraviolet? Ultraviolet is a form of light, of a kind invisible to human vision. It can be thought of as a particular "color" of light for which the human eye is insensitive (usually), though it is to some extent visible to certain birds and some insects. The important distinction between ordinary visible light and ultraviolet light lies in the difference in wavelength characteristic of each. Light, whether visible or invisible, can be thought of as being a wave motion imposed on electrical and magnetic fields, and such waves have a measurable wavelength. The waves or ripples which can be seen on the surface of a body of water have a wavelength which is simply the distance between one ripple and the next; similarly, the wavelength of light is the distance between one electric or magnetic field peak and the next. The wavelength of light can be measured by an optical instrument called an interferometer. The results of such measurements show that the wavelength of light is exceedingly small. For example, the wavelength of green light is approximately twenty-millionths of an inch, which is to say that in green light about 50,000 ripples of the field can be found over the distance of one inch.

It is useful at this point to shift discussion of wavelength measurement to a scale especially applicable to the measurement of the wavelength of light. Just as it is most convenient to measure the distance between towns in miles, the length of a block within the town in feet, and the height of a curb in inches, it is desirable to use a scale of measurement for light which produces easily handled numbers. Physicists have developed such a scale. This is called the "angstrom" scale, named after a scientist of the nineteenth century. Using the angstrom scale, the green light referred to above will have a wavelength of 5000 angstroms. The word "angstrom" is usually abbreviated to "A," so that this wavelength is usually written as 5000A.

Light can be of a single wavelength, but the light we ordinarily see will be made up of a mixture of wavelengths. The spectrograph, shown in Figure 2-1, is a convenient instrument for decomposing light into its constituent wavelengths. The essential elements of the spectrograph include a prism or grating to decompose the light and a screen upon which the light is then projected. Light whose wavelength composition is to be analyzed is directed into the spectrograph. Each different wavelength which may be contained in the entering light beam will be projected by the prism onto a different part of the screen. Light projected in this way is called a "spectrum."

The screen upon which the light is projected will be marked off in angstrom units so that the wavelength may be read off directly. If sunlight is beamed through the spectrograph, an astonishing spread of light will be projected upon the screen in all of the colors of the rainbow, and the entire region between about 3800 and 7800A of the screen will appear illuminated (Figure 2-2). Blue light will be seen at about the 4500A mark on the screen, which indicates that this is the wavelength of blue light. Green will appear at a larger value of wavelength, that is, a longer wavelength. This will be followed by yellow, orange, and red at longer and longer wavelengths. This range between 3800 and 7800A contains the wavelengths detectable by the vision of most people.

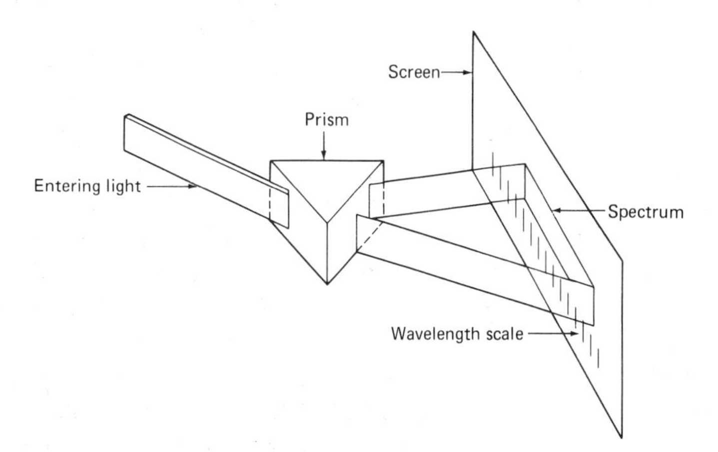

Figure 2-1. A spectrograph contains a prism which decomposes entering light into its constituent wavelengths and projects light of each different wavelength on a different portion of the screen. The screen is ruled so that the wavelength of projected light may be directly read. A quartz prism would be required if short wave ultraviolet is to be examined. A spectroscope differs from a spectrograph in that it contains optics which allow direct view into the prism so that the spectrum is seen directly, rather than by projection on a screen.

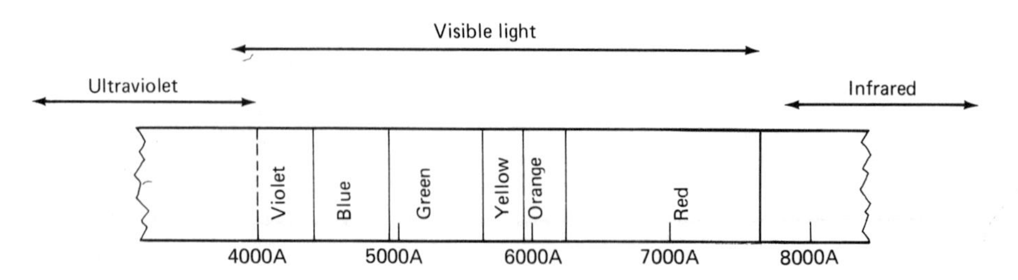

Figure 2-2. The visible portion of the light spectrum, Some ultraviolet and infrared will be projected even by a spectrograph intended for visible light.

To the right of the 7800A location the screen will be dark. Nothing will be seen in this portion of the spectrum. Yet a form of light energy is being projected on that part of the screen also. This can be confirmed by holding a thermometer in that location. The thermometer will show a temperature rise due to the presence of radiation. This is the infrared region of the light spectrum. While it is not visible to human vision, some desert snakes can detect infrared, and these snakes use this capability to home in on prey at night.

We now consider the region to the left of the 3800A mark on the screen – a region where, again, the screen is dark and nothing is seen. Once again the thermometer can be used to confirm the presence of radiation. Alternatively, a piece of fluorite held there will burst into visible fluorescence, and a photographic plate held there will be quickly darkened. This is the ultraviolet region of the light spectrum. It is this radiation which produces the startling variety of fluorescences in minerals.

Just as the visible portion of the light spectrum is divided into a number of subregions (that is, colors), ultraviolet is divided into several regions which differ somewhat in their properties. These regions, shown in Figure 2-3, are as follows. The "near" ultraviolet includes those wavelengths between 3000 and 3800A. This is what fluorescent mineral collectors call "long wave" ultraviolet since most of the radiation produced by long wave lamps is contained in this region. "Far" ultraviolet lies between 2000 and 3000A, and corresponds to the mineral collectors' "short wave" since radiation produced by short wave lamps is centered between these limits. The ultraviolet region of the spectrum continues further to the left, to wavelengths shorter than 2000A, into what is called the "extreme" ultraviolet. Extreme ultraviolet is of little interest to the mineral collector for a number of practical reasons. It is very hard to generate with any useful intensity, and further, a filter would be needed to screen out the visible light which is also produced. A practical filter capable of passing extreme ultraviolet is not available. Also, extreme ultraviolet is extremely reactive with the oxygen in the air. The oxygen molecule is decomposed by extreme ultraviolet and the ultraviolet is absorbed

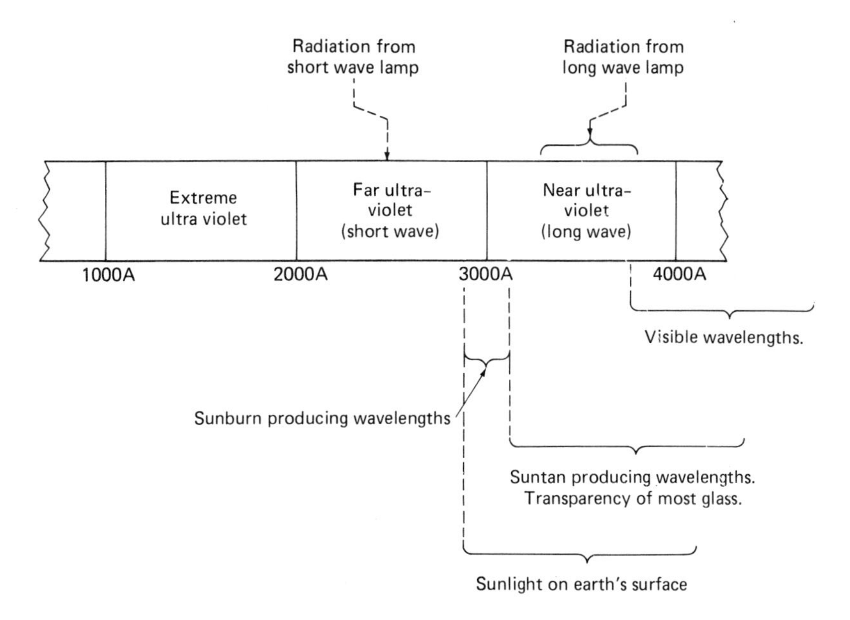

Figure 2-3. The ultraviolet region is divided into subregions. These are the long wave band near the visible blue and violet, the short wave band, and the extreme ultraviolet band. The first two bands are used to produce fluorescence in minerals. Long wave ultraviolet will pass through some ordinary glasses. Short wave tubes and filters require special glasses.

in the process, which means that extreme ultraviolet can travel only a short distance through the air before being completely absorbed. As a result, it is exceptionally difficult and expensive to construct a filtered extreme ultraviolet light, and it would be equally difficult to exploit such a light for fluorescent mineral work.

It may be wondered what novel and spectacular fluorescent effects in minerals would be produced if only a practical extreme ultraviolet light existed. For the present, the study of mineral fluorescence is limited by the existence of lamps operating at near and far ultraviolet, that is, long and short wave lamps.

ULTRAVIOLET FROM THE SUN

The sun produces substantial amounts of near ultraviolet, a good deal of far ultraviolet, and some extreme ultraviolet. The makeup of the radiation which reaches the earth's surface is quite different, however, because the light produced by the sun must first pass through the earth's atmosphere. The extreme ultraviolet is quickly absorbed by the oxygen in the upper atmosphere. Oxygen atoms, freed from the oxygen molecule by the impact of extreme ultraviolet, then recombine to form ozone. This ozone, concentrated at about 15 miles above the earth's surface has a profound effect on the transmission of far ultraviolet emanating from the sun. While ordinary oxygen is reasonably transparent to far ultraviolet from the sun, ozone is very opaque to this radiation. The ozone layer almost completely screens out the far ultraviolet from the earth's surface so that no radiation of wavelength shorter than 2900A ordinarily reaches the earth's surface.

It is very fortunate that neither extreme ultraviolet nor much far ultraviolet produced by the sun reached the earth's surface, because both forms of radiation would be deadly to life in the concentration and intensity which would exist in the absence of this screening effect. The ozone responsible for this screening exists in very low concentrations, equivalent to a layer only about one-eighth of an inch thick if compressed to the density of air at sea level. Because of this low density, the protective ozone screen is fragile and could be extinguished by even low concentrations of certain pollutants which might be carried to the upper regions of the atmosphere where the ozone exists. It is for this reason that there is so much concern about the release of fluorocarbons into the atmosphere.

Suntan is produced by ultraviolet of wavelength between about 3200 and 4000A. Tanning of the skin is a protective measure which appears to screen the skin from deeper penetration of ultraviolet. Sunburn is produced primarily by ultraviolet between about 2900 and 3100A. It is fortunate that the wavelengths responsible for tanning and sunburn are in part nonoverlapping. Effective suntan lotions take advantage of this by screening out the ultraviolet wavelengths shorter than 3100A which are responsible for sunburn, while passing the longer wavelength tanning rays. As a result, the suntan lotion promotes tanning, not by increasing the speed of tanning but by allowing longer exposure to the sun before a burn results.

WHAT THE EYE SEES

Human visual sensitivity ordinarily extends down to about 3800A. However, remarkably enough, the seeing capabilities of some people, particularly some young people, extend down to almost 3100A. What color is seen at these wavelengths? For those exceptional people who can see well into the long wave ultraviolet, the resulting color is blue! What is happening is this: the cornea of these people is transparent to almost the entire long wave ultraviolet band and this ultraviolet enters the eye. Here, it causes blue and green fluorescence in certain elements of the retina of the eye. The light-detecting portions of the retina then detect this blue and green light.

Some collectors are bothered by a blue haze surrounding minerals when long wave ultraviolet is used. This is due to reflected ultraviolet producing fluorescence in the eye. For most persons, of course, this never occurs, but for those fluorescent collectors who are bothered by the blue haze produced by long wave ultraviolet lights, special goggles can be purchased which eliminate this problem.

TRANSMISSION

Most materials which are transparent to visible light are also transparent to near or long wave ultraviolet. Ordinary glass, for example, transmits from the visible band down to about 3100A, though with reduced effectiveness as shorter wavelengths are approached. With some care, glass is easily formulated with high transparency in the near ultraviolet, and such glass is employed in long wave ultraviolet mineral lamps. By contrast, very few materials are transparent to far or short wave ultraviolet, and all conventional glasses are virtually opaque at the 2537A wavelength where short wave ultraviolet lamps operate. Thus, special glasses must be used for the fabrication of short wave lamps.

Since ultraviolet lamps produce substantial amounts of visible light, the output of the lamp must be passed through a filter capable of blocking visible light while passing ultraviolet. Such filters are difficult to produce in the short wave regions since they must be much more transparent to short wave than to visible light. The development of a successful short wave filter was one of the important steps which made the collection and study of fluorescent minerals possible. The development of the modern ultraviolet mineral light is traced in the next chapter.

3
Ultraviolet Lights for Mineral Fluorescence

Alchemists stumbled upon minerals or mineral preparations which, when held for a time in the sun and then quickly taken indoors, would continue to glow. The long wave ultraviolet and the visible violet produced by the sun were the sources of excitation of this phosphorescent response. A few minerals fluoresce with such intensity that the effect is noticeable in sunlight. Thus, some willemites owe their intense green color in daylight to the fluorescence produced by solar ultraviolet; some fluorites, their blue; and some rubies, their red in part to this effect.

This type of excitation is obviously unsatisfactory for the general study of fluorescence, since in most fluorescent minerals any such effect produced is usually subtle and easily overwhelmed by the visible light from the sun which reflects from the specimen. However, in the early decades of this century, some collectors took advantage of the sun as a long wave light source. A light box was rigged with a window which was covered by a filter glass. The filter passed long wave ultraviolet into the box while filtering out most of the visible light. A mineral could be placed in the box and the fluorescence, if any, viewed through a small opening. This too is unsatisfactory for the general study of fluorescence, as it cannot be used indoors or at night. Further, no short wave can be provided in this way, since none of the sun's short wave ultraviolet radiation can penetrate the earth's atmosphere.

An ordinary incandescent (hot filament) light bulb produces no short wave ultraviolet, a small amount of long wave ultraviolet, a large amount of visible light, and an even larger amount of infrared radiation. Such a light can be used, in conjunction with a filter which removes most of the visible light, as a source of long wave ultraviolet, but the amount produced is so small that only a very few minerals will fluoresce, and then only feebly. Special incandescent lamps are sold with bulbs of a very dense violet filter glass for the purpose of illuminating fluorescent posters. In such bulbs, the glass gets excessively hot and produces only negligible long wave ultraviolet.

Thus, our two most familiar sources of illumination, the sun and the light bulb, are poor sources of the ultraviolet needed for the generation of fluorescence in minerals. The entire problem faced in earlier years was to find sources which produced a plentiful amount of filtered ultraviolet while producing only small amounts of visible light and heat. Further, the whole arrangement had to be easy to operate, dependable, long-lived, and capable of being sold at a cost which the mineral hobbyist could afford.

It took several decades for solutions to these problems to emerge, culminating in modern ultraviolet lights. It is useful to review the earlier developments before discussing those available at present.

EARLY ULTRAVIOLET SOURCES

The iron arc or spark, a copious source of short wave ultraviolet and to a lesser extent of long wave ultraviolet, was perhaps the most widely used source of ultraviolet light for mineral fluorescence work early in this century. It was used as early as 1903 in one of the most extensive investigations of mineral fluorescence ever undertaken. The investigation by Kunz and Baskerville tested some 13,000 different mineral specimens in the American Museum of Natural History. Numerous mineral fluorescences were discovered in this investigation. Many more would undoubtedly have been found if a good filter had been available at that time to remove the visible component of the light produced by the spark.

The iron arc was also used by Charles Palache in his investigations of the fluorescence of Franklin minerals, and in a paper in 1928 he described the fluorescence of Franklin calcite, calcium-larsenite (esperite), pectolite, margarosanite, hardystonite, hedyphane and, of course, willemite.

Palache probably got the idea for this investigation from the even earlier use of the iron arc at the mines at Franklin. As early as the first years of the 1920s, in what may have been the earliest industrial use of mineral fluorescence, the iron arc was used to examine mill tailings for willemite, as well as for general assay work. One mineralogist who visited the mines at that time wrote:

> In several of the laboratories and offices there was fixed to the wall in a dark corner a small apparatus giving a high tension spark which could be switched on from the lighting circuit. A piece of ore held beneath the spark showed up any willemite present by a vivid green fluorescence. Some very pretty effects were obtained; for example, specks of pale green willemite embedded in snow-white calcite glowed as brilliant green spots in a crimson background.*

Later, the iron arc was also employed at Franklin for exploration within the mine and was kept handy to the sorting table to aid in mineral identification.

---

*Spencer, G. J. 1929. Fluorescence of minerals in ultraviolet rays. *Journal of the Minalogical Society of America* **14** (1):33–37.

Figure 3-1. John Baum, after whom the fluorescent mineral johnbaumite is named, positions a mineral specimen under an antique iron arc lamp, the kind once in frequent use in the mines and offices at Franklin, New Jersey.

The early iron arc consisted simply of an open gap between iron bolts supplied with 9000 volts from a 250-watt step-up transformer. Similar ones were home built by collectors of fluorescent minerals. Even portable battery operated units were rigged by early enthusiasts, at a much lower power level of course. The color effects produced are indeed vivid and, as the authors of one article commented:

> Such a variety of colors gives a truly kaleidoscopic effect to certain specimens, and thus it is to be expected that such specimens are eagerly sought after by collectors for it is to be doubted if anything equal in beauty is to be found anywhere else in the world.*

---

*Gunnell, E. M. and Shrader, J. S. 1935. New Jersey willemites show spectacular fluorescence. *The Mineralogist* 3 (1):9–10, 22.

The Gerstmann Museum and the Mineral Museum in Franklin both preserve an iron arc in working condition.

During the 1920s and 1930s, one of the most popular sources of long wave ultraviolet was the argon bulb. This was a lamp which looked much like an ordinary incandescent bulb and, like the ordinary incandescent lamp, could be used in an ordinary screw socket. Within the lamp, in place of a filiment there were two metal plates or electrodes separated by a small gap. The bulb was clear glass and the inside contained argon gas. When the bulb was turned on, an electric glow arc was formed between the electrodes, and the electrified argon gas emitted a useable amount of long wave ultraviolet. Only a small amount of visible blue light was produced, and this lamp could be used without a filter. Ultraviolet output was weak, but because of the low cost of this bulb, a bank of bulbs could be used to light up a display cabinet of fluorescent minerals. The argon bulb was also used directly with 90-volt batteries for field work.

For museum displays requiring copious amounts of long wave ultraviolet, the Nico lamp was used. Available as early as 1920, this lamp outwardly resembled a present day fluorescent light. It consisted of a long glass tube containing mercury. An electric arc was struck by tilting the tube so as to allow liquid mercury to act as one electrode within the tube. Like a modern "black light" tube, the glass envelope was made of filter glass which removed much of the visible light. This filter was based on nickel and cobalt (thus Nico) in a selected glass material. This filter was suitable for long wave lamps only, however.

In the 1930s and early 1940s, a variety of lamps were made of fused quartz tubing containing mercury in which an arc could be struck by high voltage to produce short wave ultraviolet. Some of these were constructed in a spiral or similar compact shape in order to concentrate the ultraviolet produced into a small source area. This increased the intensity of short wave ultraviolet which could be directed at a specimen and made it convenient to use a lesser amount of short wave filter glass which was beginning to become available.

Most of the ultraviolet light sources described above were either expensive, short on working life, difficult to operate, low on ultraviolet output, or excessive in the amount of visible light produced.

There were other developments of the 1920s and particularly the 1930s which set the stage for the ultraviolet lights used by mineral hobbyists today. There were: the development by Corning in the 1920s of a filter glass suitable for short wave ultraviolet, the development of high pressure mercury arc lamps for street and industrial lighting and for suntanning, the development of low pressure mercury arc sterilizing lamps, and the development of fluorescent lighting. None of these lamp developments were directly aimed at the

Figure 3-2. One of the earliest ultraviolet lights made specifically for fluorescent mineral work, V43 was manufactured by Ultra-Violet Products, Inc., between 1943 and 1945. Both line powered and battery powered portable versions were produced. With visible light absorbing filter removed, the photo shows a serpentine shaped ultraviolet tube. This light is still to be found in use in the offices of professional mineralogists and geologists.

benefit of the fluorescent mineral collector, but the collector substantially benefited by these developments nonetheless.

PRESENT DAY LIGHTS

The source of ultraviolet radiation is referred to as an ultraviolet lamp. When a suitable lamp is mounted in a fixture and housing, and combined with an appropriate filter, switch, ballast, and other components to make a self-contained assembly, the result will be referred to here as an ultraviolet light. The starting point in a description of presently available ultraviolet lights is the generator of ultraviolet, the lamp.

All present day ultraviolet sources used in the study of fluorescent minerals are based on the mercury vapor arc lamp in one form or another. An electrical discharge between separated electrodes in an atmosphere of mercury vapor is a prodigious producer of ultraviolet radiation, and the design of such lamps can be tailored to produce predominantly short wave, long wave, or even visible light for illumination purposes. The radiation from the mercury arc is made up of a number of spectral bands or spectral "lines." They are termed "lines" since this is what they look like through the spectroscope. While many lines are produced, the strongest set includes three lines in the ultraviolet, one of which is short wave and the other two long wave, and four lines in the visible violet, blue, green, and orange portions of the spectrum.

The energy produced in the various spectral regions can be adjusted so that either the short wave line, the long wave lines, or the visible portion dominates. This is accomplished by designing different types of mercury arcs to operate at different gas pressures and temperatures. Very low pressure lamps produce short wave radiation predominantly. High pressure lamps produce little short wave, but large amounts of long wave and visible radiation. Very high pressure mercury arcs produce output mostly in the visible region of the spectrum. The low pressure lamp and the high pressure lamp are the bases of lights of interest to the fluorescent collector.

Low Pressure Short Wave Lamps

Developments in the 1930s produced the familiar fluorescent* tube lamp used as an economical source of lighting at home, in office buildings, and in industry. At the same time, a series of germicidal or sterilizing lamps were developed making use of the same design principles, the main difference between the fluorescent and germicidal lamps being that the germicidal lamp uses a tube which is uncoated and transparent to short wave ultraviolet as will be explained shortly.

A short wave lamp is identical to a germicidal lamp. It consists of a glass tube with a coiled electrical filament at each end. Two contact pins connected to the filaments exit the tube at each end. The entire tube looks exactly like a fluorescent tube. In fact, the dimensions, components, and method of construction are similar to that of fluorescent tubes. The interior is filled with argon gas and mercury vapor at low pressure. When the lamp is turned on, a current is first passed through the filaments. The heated filaments strip electrons free from the argon atoms. These free electrons make the gas in the tube an electrical conductor. Now a high voltage is applied from one end of the tube to the other, which accelerates the electrons down the tube. These electrons impact the mercury atoms and transfer energy to electrons of these atoms, in response to which the elec-

*"Fluorescent" tube lamps do not produce fluorescence in minerals. Their purpose is to provide visible light for illumination purposes. They are called "fluorescent" because the visible light is produced by an artificial fluorescent material, or phosphor, which coats the inside surface of the tube.

trons move to higher orbits. On returning to their original orbits, the electrons radiate large amounts of 2537A short wave radiation, and lesser amounts at other wavelengths.

The low pressure and the associated low temperature of operation are essential for the dominance of the 2537A output. This can be seen in Table 3-1 which compares the spectrum from the high pressure mercury arc bulb to be discussed shortly with the output from a low pressure short wave lamp. Note that the low pressure lamp produces 86% of its output in the 2537A short wave line. By contrast, the high pressure bulb produces only a little at this wavelength, a fairly high amount at 3650A (long wave), and a substantial amount of visible light. After filtering, a different balance results, which will be discussed later.

Unlike the household fluorescent lamp, the germicidal or short wave lamp is not made of ordinary glass since glass is not transparent to short wave radiation. Fused quartz is very suitable in terms of transparency and will transmit approximately 95% of the radiation at 2537A in the thickness commonly used in lamps. However, quartz fuses only at a high temperature and is difficult to work with in terms of the fabrication methods ordinarily used in lamp manufacture. Certain special glasses with high transparency at 2537A are therefore used. Pyrex No. 9741 with a transparency of about 75% and Vycor No. 791 with a transparency of about 85% are examples of such glasses. The need to use these special glasses means that the cost of a short wave lamp will be substantially greater than that of an ordinary fluorescent lamp of identical dimensions, but less than that if quartz were used. If it were not that the food industry, hospitals, and the

Table 3-1. Relative Strength of Mercury Spectral Lines (Unfiltered).

1 Color	2 Wavelength	3 Low Pressure (SW Tube)	4 High Pressure (LW Bulb)
(Invisible; SW ultraviolet)	2537A	86%	5%
(Invisible; LW ultraviolet)	3200A	2%	10%
(Invisible; LW ultraviolet)	3650A	2%	19%
Violet	4047A	2%	11%
Blue	4358A	5%	18%
Green	5461A	3%	24%
Yellow	5781A	0%	13%

like use these same short wave lamps to maintain sterile conditions, thus creating an additional market base for such lamps, the price for mineral use would be materially higher.

Low Pressure Long Wave Lamps

The majority of long wave light sources in use by mineral collectors are based on a design similar to that of the low pressure short wave tube described above. The long wave tube is identical in size, shape, and operating characteristics to the short wave tube.

Somewhat paradoxically, the generation of long wave begins with the generation of 2537A short wave, precisely as described earlier. In the long wave tube, however, the inside surface of the tube is painted with a white substance, a phosphor. When the short wave generated within the lamp strikes this phosphor, it is absorbed and converted to long wave ultraviolet radiation. This is a fluorescence process, even though the output is primarily in the invisible long wave ultraviolet region rather than in the visible light region. Such phosphors are synthetic fluorescent minerals tailored to this purpose. Typically, they consist of a synthetic apatite activated with thallium or cerium. The ordinary "fluorescent" light with which we are all familiar as a means for lighting in the home or office is based on exactly the same design. The only difference is that the phosphor is selected to produce its main output in the visible region of the spectrum rather than in the long wave ultraviolet. It is thus clear how modern ultraviolet lamps, both short and long wave, have benefited from research directed to the development of fluorescent lights and also from the sharing of parts, components, manufacturing facilities, and techniques.

High Pressure Long Wave Lamps

The high pressure mercury arc bulb is only occasionally used in long wave mineral lights, since it is expensive, slow to warm to working temperature, and hot in operation. Its advantage is that it can produce several times the ultraviolet power obtainable from an ordinary hand held long wave lamp of the more conventional kind. One such bulb is available under the designation H100PSP44-4, operating at 100 watts. This lamp looks like an ordinary spotlight. However, inside the bulb there is a fused quartz tube containing mercury at eight atmospheres pressure. An electrode at each end of the quartz tube provides the electric arc.

At this high pressure, the mercury vapor arc produces substantial output centered on the 3650A long wave line of the mercury vapor

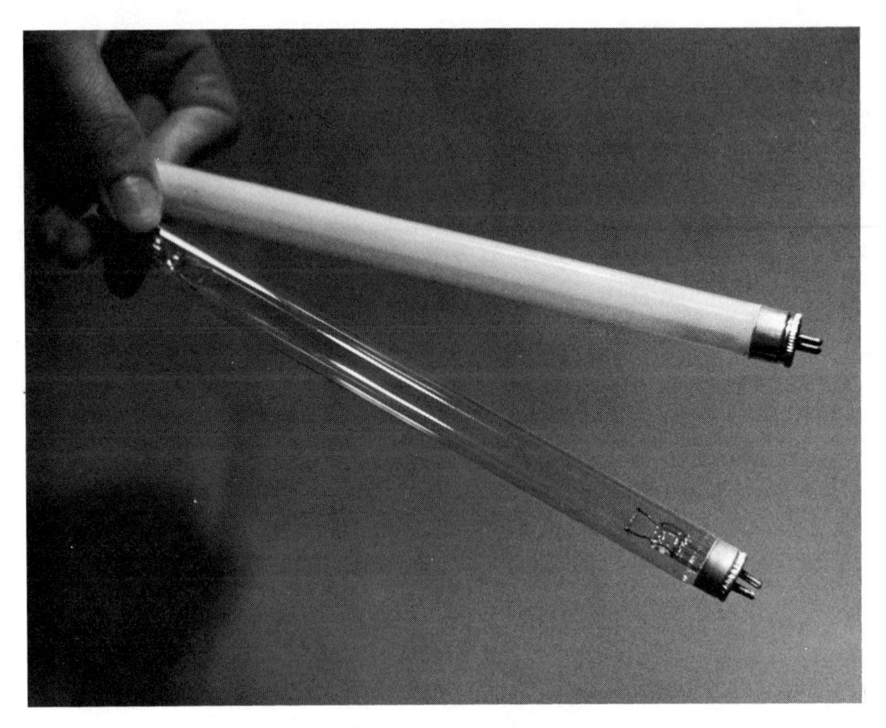

Figure 3-3. Two ultraviolet tubes used in the Raytech Industries LS-88. This ultraviolet light produces either long wave or short wave ultraviolet, depending on which of the two tubes is turned on. The white tube is internally coated with a long wave ultraviolet producing phosphor and is quite similar to a "fluorescent" light tube of the same size. The clear tube is uncoated and made of glass with a high short wave ultraviolet transparency, allowing the short wave ultraviolet produced by the electrical excitation of mercury vapor to pass. Such tubes have almost completely replaced the specially fabricated tubes of the kind shown in Figure 3-2.

spectrum and little short wave radiation. The short wave radiation that is produced is further reduced by the outer glass envelope which is a poor transmitter of short wave.

A substantial amount of visible light is produced by such bulbs, and as a result, for fluorescence work, an external filter must be used which is capable of passing 3650A but will absorb visible light. Since a filter operated in conjunction with a high pressure lamp will get hot, it must be made of a heat resistant glass. Corning produces suitable filters, filter 5840 being an example. Column 4 of Table 3-1 shows the relative output of a high pressure lamp before filtering. After filtering by 5840 glass, the 3650A line is reduced to 10%, the 3200A to 1%, all others to substantially zero.

Several other bulbs are generally similar in their output characteristics, such as the H85A3/UV, the H100A4/T, and the H100A38 - 4, but are built into bulbs of various shapes other than spot and so must be used with a suitable external reflector to concentrate the ultraviolet output on a small area. Complete lights, including filter, have been available from one or another of the companies that supply ultraviolet apparatus to the mineral collector.

While such lamps are of interest to the serious collector, the type of long wave lamp ordinarily used and ordinarily found in the form of hand held, display, or portable battery operated lamp is based on the low pressure lamp described earlier.

Filters

For effective use in the study of fluorescent minerals, the radiation from the ultraviolet lamp must be filtered to remove the visible light which accompanies ultraviolet generation. Most fluorescent minerals are white or light in color, and thus would reflect appreciable amounts of the visible light produced by the lamp. By contrast, the fluorescent conversion of ultraviolet to visible light by fluorescent minerals is usually very inefficient, resulting in only a low level of output intensity. The result is that, without a filter, reflected light will overwhelm the visible light produced by fluorescence in all but the most intensely fluorescing minerals. A suitable filter is thus an indispensable component of an ultraviolet light for mineral fluorescence.

The development of a suitable filter for short wave ultraviolet is a significantly greater problem than that of a long wave filter. While it is easy to produce a glass that will transmit long wave, it is more difficult to produce one which will pass short wave. A short wave filter is even more difficult, since coloring agents must be introduced into the glass which will block visible radiation while leaving short wave virtually unaffected. It was not until the 1920's that Corning introduced such a glass, and it was this filter glass, together with the tube developments described above, that made our present day short wave mineral lights practical. While ordinary glasses utilize silica to form a silicate based melt, the short wave filter is based on a phosphate melt. Phosphates can also form glasses, as witness the "phosphate bead," a clear glass sphere formed on a platinum wire of a cooled phosphate melt, used to test for the presence of various metals in minerals. Such phosphate glasses are highly transparent to ultraviolet radiation. They are also excellent solvents of metallic oxides as shown by the coloration of the bead after it is touched to a metal oxide when hot. It is a balance of selected metal oxides which produces the desired short wave filtering effects, passing short wave ultraviolet while retarding visible light.

The manufacture of such filters calls for the most careful control of conditions. New, scrupulously clean crucibles are used for each melt; ingredients are carefully measured out; and melting is done under conditions which exclude air, since traces of iron or titanium which may be present must be in a low oxidation state. The melt is then poured over a flat surface and smoothed, after which it is cut into sections and may then be polished. Finally, each batch is tested to determine if it meets the manufacturer's specifications for ultraviolet transparency and visible light stopping power. Not every batch passes.

Since rejection of batches is costly, since the special care taken throughout manufacture of the filter glass is also costly, and because the market base is small (the germicidal application of short wave light does not require a filter), the cost of a short wave filter is high.

The Corning designation for their short wave filter is glass No. 9863, usually sold in 3-mm thickness. Schott produces an apparently similar filter in Germany under the designation UG 5; Chance in Britain under the designation OX 7; and Hoya in Japan under the designation U-330. These have a useful transmission only between about 2400 and 4000A, and transmit with only about 30–60% efficiency at 2537A. This somewhat low transparency to short wave radiation is one of the penalties intrinsic to the design of such filters.

Another penalty is the deterioration or aging which these filters experience, a process called "solarization," which is the deterioration of ultraviolet transparency after some time in use. However, measurements made on the filters recently developed by Hoya in collaboration with Ultra-Violet Products, Inc., appear to indicate that these filters are substantially immune from the deteriorating effects of solarization. Whatever the ultimate cause of filter solarization may be, Hoya appears to have successfully stabilized its filters against it. This represents an important improvement in the short wave filter, perhaps the most important since the original short wave filter development by Corning.

Short wave filters are slightly transparent in the visible violet and blue regions. Since the mercury lamp generates a weak violet spectral line, the filtered light appears to the eye as a weak violet-blue color. This visible light leakage can cause confusion in examining some fluorescent minerals, hardystonite for example, whose fluorescent color closely resembles this visible leakage. The filter is also slightly transparent in the red region of the light spectrum. Around the tube filaments, a certain amount of red glow is produced by argon gas which is contained in the tube, and as a consequence of the slight filter transparency, a small amount of visible red light is also radiated. This can cause some confusion in examining reddish or brownish materials, as the reflected red light can be mistaken for a weak red fluorescence.

The long wave radiation produced by phosphor-coated long wave tubes covers a broad band, primarily between 3000 and 4000A, with the peak output around 3600A. About 3% of the total radiation is in the visible region; consequently, as with the short wave tube dis-

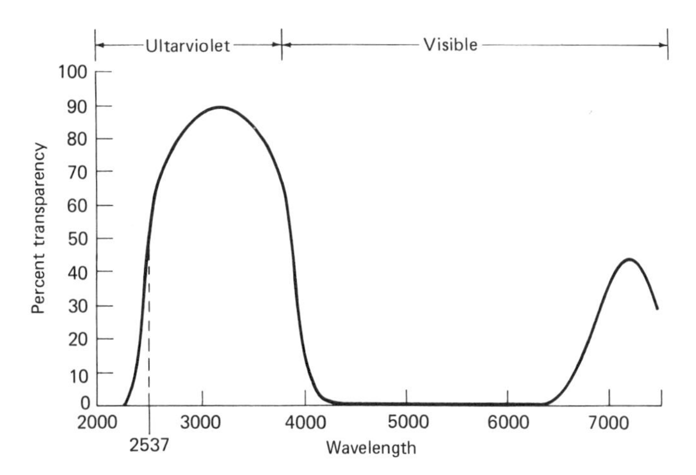

Figure 3-4. Transparency curve for a short wave filter. Between approximately 4200A and 6500A, the filter is opaque. Thus, the blue, green, and yellow spectral lines of the mercury lamp are almost blocked; but the violet line at 4047A is passed, as is red light around 7000A produced near the lamp filaments. The desired short wave spectral line at 2537A is passed by the filter, but at 50% or less transparency.

cussed earlier, filtering is required for fluorescent mineral work. Two approaches to filtering are taken with such lamps. In the first, a separate filter glass is used as with short wave lamps. Corning glasses Nos. 5840, 5970, or 9863 will do – selection depending on price or availability. Some small amount of visible violet or blue will penetrate these filters, and as with short wave lights, some confusion with visible violet fluorescence can result. In the second approach, the glass from which the long wave lamp is made is colored so as to constitute a built-in filter. Lamps built in this way are popularly used to

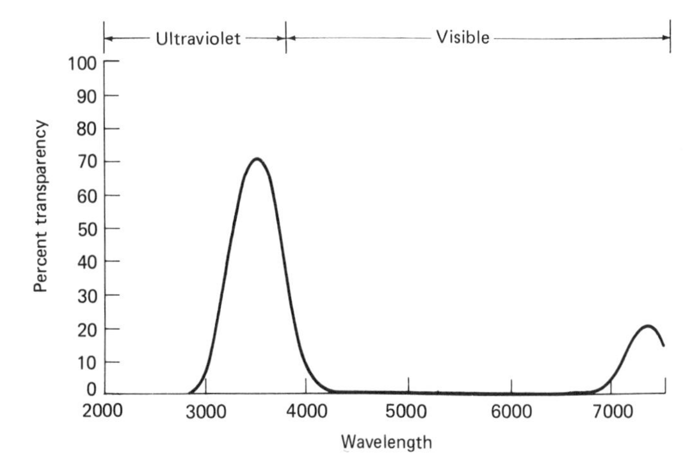

Figure 3-5. Transparency curve of a long wave filter. The broad band phosphor used to produce long wave ultraviolet in low pressure lamps typically provides maximum intensity at about 3600A, close to the maximum transparency region of the long wave filter, but the phosphor also produces some light in the violet region which the filter passes. Red light near 7000A produced near the lamp filaments is also passed by the filter.

illuminate fluorescent posters and are often called "black light" tubes.

Generally, lamps produced with separate filters provide superior filtering and are preferred for mineral study. On the other hand, self-filtered black light tubes are significantly less expensive. Some variation in fluorescent mineral response can be seen between various brands of black light tubes, probably due to differing phosphor compositions or differing built-in filter characteristics, and between these and long wave lamps with separate filters. Also, the fluorescent mineral response produced by the high pressure bulb described earlier can be quite different, at least for a few minerals. Thus, a Franklin barite may fluoresce pale yellow under the lamp with separate filter, but may not fluoresce under the black light tube or high pressure bulb.

SELECTING ULTRAVIOLET EQUIPMENT

Ultraviolet lights in a variety of sizes and power levels are available on the market, exploiting the short and long wave lamps, and filters just described. These are sold by companies specializing in the manufacture of ultraviolet equipment. Two names presently dominate this field: Ultra-Violet Products, Inc., of San Gabriel, California, and Raytech Industries, Inc., of Stafford Park, Connecticut. Both companies produce a full line of ultraviolet lights for mineral fluorescence. These include equipment in the three categories: hand held lights, portable field lights, and show or display lights. Ultra-Violet Products favors tough plastics as the housing for its hand held and portable lights. Raytech Industries uses metal housings for most of its products. Raytech offers the greatest variety of dual band lights, while Ultra-Violet Products has introduced long-lived Hoya-type filters into all of their short wave units. There is thus ample choice, perhaps to the point of confusion.

To obtain a better understanding and comparison of the lights available, a useful starting point is to consider the types of ultraviolet generating lamps upon which these various lights are based. These are primarily low pressure tubes of the kind discussed earlier, which are built like a fluorescent lighting tube but contain an ultraviolet phosphor for long wave or an uncoated tube of identical size and shape but of special glass for short wave.

Table 3-2 shows the sizes available from major electric lamp manufacturers in terms of power rating and tube length. The serial number identification is also shown. Column 4 checks off long wave lamps for which a separate filter is needed. Column 5 checks off long wave lamps with filter capabilities built into the glass, so that no external filter is needed. These are the so-called black light tubes. The last

Table 3-2. Lamp Characteristics.

1 Power (watts)	2 Length (in.)	3 Designation	4 LW Filter Needed	5 LW Filter Built-in	6 SW Filter Needed
4	5 1/4	F4T4 BL*	X		
		G4T4*			X
4	6	F4T5/BL	X		
		F4T5/BLB		X	
		G4T5			X
6	9	F6T5/BL	X		
		F6T5/BLB		X	
		G6T5			X
8	12	F8T5/BL	X		
		F8T5/BLB		X	
		G8T5			X
15	18	F15T8/BL	X		
		F15T8/BLB		X	
		G15T8			X
20	24	F20T12/BL	X		
		F20T12/BLB		X	
25	18	G25T8			X
40	48	F40BL	X		
		F40BLB		X	
85	22	F72T12/BL/HO	X		

*These two tubes are U-shaped. All others are straight tubes.

column checks off the short wave tubes for which an external filter is, of course, needed.

As is evident from this table, two or sometimes three different types of tube may be designed with the same power rating, length, diameter, and end fittings. This allows the manufacturer of an ultraviolet mineral light to provide a number of different lights by outfitting a limited number of equipment designs with tubes differing in ultraviolet characteristics but identical in external design. The flexibility is increased further as the manufacturer combines tubes in various ways. In addition to the tubes shown in Table 3-2, manufacturers of mineral lights have arranged with tube manufacturers to supply special tubes, partly phosphored as described below, to provide dual band output from a single tube. This further increases the variety of lights which can be, and indeed are, offered.

The 4- and 6-watt tubes are most frequently used in conjunction with separate filters in the form of hand held mineral lights operated from ordinary electric outlets. For this purpose, one or two tubes may be built into a light. In those with two tubes, they may both be short wave or both long wave, or one of each may be included to form a dual band light. If the light contains both a short and a long wave tube, a short wave filter is used since this is satisfactory for both bands. In some lights, either the short or long wave may be turned on separately, while in others both are turned on at once. In yet another type of dual band light, one tube is used, coated with long wave phosphor over half its length, uncoated and short wave transparent over the other, so that it radiates long wave from one end and short wave ultraviolet from the other at the same time. Such split function tubes are usually used in conjunction with two filter sections: a long wave filter covers the long wave half of the tube, while the other half is covered with a short wave filter.

The hand held light is the minimum essential light which a fluorescent collector must have and is usually the one with which he begins. It is best if one can afford it to have one of each: one hand held short wave light and one long wave light. Separate lights are generally superior to dual band lights since they will usually produce more ultraviolet power in a given band than will the dual light. In other regards, a dual band light in which one band may be turned on at a time is satisfactory. Somewhat less desirable for most purposes is the light utilizing the type of tube coated at one end only. This allows short wave to be produced at one end and long wave at the other; as a consequence, one can block out one band or the other by holding a hand over half the light. By this means, it will be possible to determine if a specimen is a short or long wave fluorescent and to see also the totally different responses which some fluorescent minerals display under short and long wave.

While separate lights, or at least lights in which the two bands can be separately directed at a specimen are preferable, the type of light in which both bands are on at once has certain uses. When scanning minerals at a show or examining the dealer's stock in his shop, it is inconvenient to first examine a large number of specimens under one band and then to repeat the process over an extensive stock with the other. The process is tiring – tiring to the collector and tiring also to the restless dealer who is concerned about so much of his material disappearing under cloth, table, or into a closet for so long. The dual band light in which both bands are on at once is very useful here, as a fluorescent mineral can be spotted quickly and its response under different bands determined later.

Portable field lights are based on the same 4- or 6-watt lamps or tubes used in hand held plug-in lights. Like the hand held lights, these may be found in one band or dual band versions. Modern portable ultraviolet lights are based on the same principle as the various popular camp lanterns. These utilize a transistor chopper or oscillator and a step-up transformer, and operate from two 6-volt lantern batteries. Such lights are very efficient, and many hours can be got-

Figure 3-6. A collection of ultraviolet lights produced by Ultra-Violet Products, Inc. The large black unit at the left is a long wave light designated B-100A and based on a high pressure bulb. The remaining lights are based on low pressure mercury tubes. From the left counter clockwise these are: the UVG-54 hand held, line powered, short wave light; two pocket size battery portables, the H4-S and H4-L, for short and long wavelength; the hand held UVG-11 line powered, short wave light; the UVG battery powered short wave portable; in the center rear is the G-14 rechargeable battery portable. The short wave lights above have long wave equivalents in an identical housing. The large portable is also made in a dual band form. (Photo specially provided by Ultra-Violet Products, Inc.)

ten from one pair of these relatively low cost batteries. Smaller, almost pocket size, portable ultraviolet lights are also available. These utilize flashlight size batteries and, of course, provide a lower ultraviolet light intensity and shorter battery life.

It is inconvenient to carry two such lights into the field. As a consequence, a dual band light is preferred for field use. Since a vast expanse of rock in a quarry, mine dump, or mine itself will usually need to be examined, it is useful to have both ultraviolet bands on at once. As a practical matter, this means that a single split tube design is preferred, since a battery source should not be expected to support two lamps at one time. However, if investigations are to be limited indefinitely to a region in which only short wave or only long wave minerals predominate, a single band portable may be preferable.

Most display or show lights exploit 15-watt or higher power tubes and, like the lights described earlier, may use one or two tubes and may be single or dual band. The collector's choice must relate to

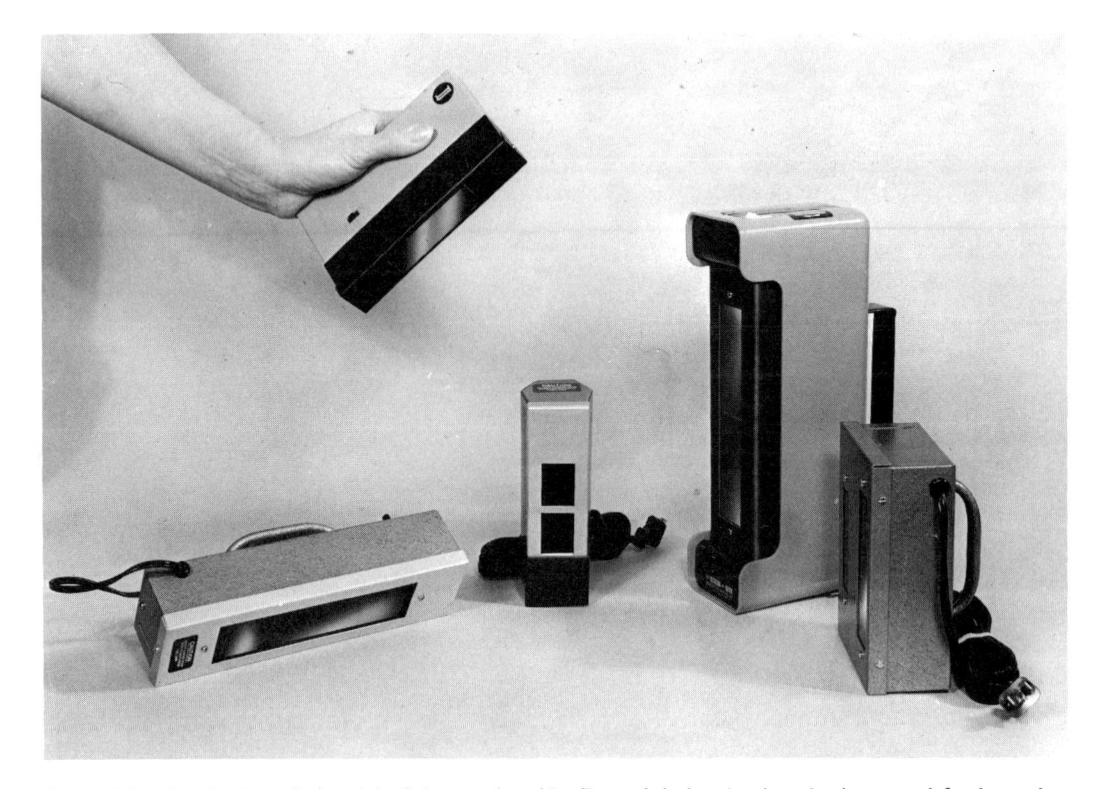

Figure 3-7. A collection of ultraviolet lights produced by Raytech Industries, Inc. In the upper left, the pocket size PPS#10A short wave battery powered portable. From left to right at the bottom are: the LS-88 dual band, hand held, line powered light; the small LS-4 dual band, hand held, line powered light; the R3-S "Raytector III" battery portable short wave; and the LS-7 dual band line powered light. The two portables have long wave equivalents in the same housing, and the Raytector III is also made as a dual band light. (Photo specially provided by Raytech Industries, Inc.)

how he will display his minerals and must be worked out in these terms. Due to the rapidity with which conventional short wave filters solarize under the high intensity used in a show or display light, the collector may prefer the newer long-life filters described earlier, particularly for the show or display application. For long wave display, a black light tube, containing a filter colorant in the glass of the tube, may provide satisfactory results at significantly lower cost than a separate tube—separate filter design.

AGING AND RESTORATION

Probably every fluorescent hobbyist has observed that under a new short wave light, minerals will fluoresce brilliantly at first. As time goes on, as months or years pass, the fluorescent response seems to wane. Some collectors may conclude that the minerals are losing their powers of fluorescence. However, as far as is known, minerals do not deteriorate in fluorescent power and can continue to fluoresce indefinitely. The decrease in fluorescent performance which is being seen is really due to the short wave light whose output of ultraviolet deteriorates as time goes on. This deterioration occurs in both the tube and the filter.

Most glasses slowly decrease in ultraviolet transparency due to continued exposure to ultraviolet. This is particularly pronounced when short wave is involved. The Pyrex 9741 glass from which most short wave tubes are made is not immune. Further, after long use, the insides of these tubes become coated with a metallic film which results from sputtering or evaporation from the hot lamp filaments, and this also contributes to a decrease of transparency. Given enough operating time, both of these effects together result in a substantial decrease of lamp output, with most of the problem attributable to deterioration of transparency in the glass itself. To restore full performance, the tube or tubes must be replaced.

Some short wave ultraviolet filters also lose transparency with use, but the cause of the problem is less certain. After some time in service, the back of a filter, that is, the side facing toward the short wave tube, will become covered with a thick, whitish film. Some believe that this is the entire source of the reduction of transparency of the filter. Others attribute the deterioration to a change which takes place internal to the glass structure itself, in effect, solarization. Whichever the cause, the net result is a decrease in ultraviolet transparency and fluorescent response.

Various filter lifetimes can be projected, but some measurements indicate a definite deterioration in short wave ultraviolet transparency for most filters after perhaps five to ten hours of use. This refers to the cumulative time that the light is actually in operation, and since a hand held ultraviolet light is usually used sparingly, a life span of several years is typical, after which something often must be done about the filter. The simplest solution is to replace it. However, short wave filters have always been expensive and are now becoming increasingly so. It would be desirable, therefore, to be able to restore these filters to useful life if possible.

Opinions differ with regard to the restorability of short wave filters, with some people holding that there is no successful method of restoring transparency, while others suggest one or another of various methods. One method claimed to be successful calls for the removal of the white film. This recommendation is tied to the view that this film is the cause of the deterioration in filter transparency. Another method calls for the filter to be raised to a high temperature for some time, a method which is known to reverse solarization in other types of glass.

The following approach has been found to completely restore a filter in one experiment. The white film, if present, was first removed with repeated buffing with a soft cloth and a Bon Ami paste. Then the filter was placed in a temperature controlled oven or kiln, and rested on one edge with the support of some ceramic blocks. The temperature was then raised slowly, over at least one hour, to 430° C. The filter was then "soaked" at this temperature for about two hours. The temperature was then slowly dropped back to room temperature, a process again taking over one hour. This very slow change in temperature was required to prevent stresses from being set up in the glass because of uneven heating. Such stresses could result in cracking to which this material is susceptible due to a high coefficient of thermal expansion. After this treatment, the transparency of the filter to short wave ultraviolet was found to be brilliantly restored. While this is not an assured solution to filter restoration, it is worth trying before a used filter is discarded.

4
Collecting Fluorescent Minerals

Good fluorescent mineral collections are the result of constant study and diligent search for superior material. As in the development of mineral collections generally, purchase and trade provide rewarding avenues for improvement. However, the fluorescent collection, perhaps more than other mineral collections, is based on active collecting in the field. Collecting in quarry, mine dump, or mine itself is an essential, indispensable activity among fluorescent collectors.

This appears strange at first sight, for the collecting must be done in the unfamiliar, and at first uncomfortable, element of darkness. Why then is it done? The chance to find one's own mineral specimens is always appealing and this may provide the initial motivation, but the experience of collecting soon provides other rewards. When the site, carefully scouted by day, is investigated with a portable ultraviolet light at night, it is found to be rich with good fluorescent material. The site may have been picked over for years by collectors working in daylight, but by night, perhaps new, as yet unfound, treasures are revealed. Fluorescent specimens, often unimpressive or uninteresting in daylight appearance, have been overlooked by daylight collectors unaware of their beauty under the ultraviolet light.

Then there is color. Before darkness has fully set in, the fluorescent color seen under the ultraviolet light is weak, uncertain, and faint even when the ultraviolet light is held close. As it continues to grow dark, a moment is finally reached which, in the Middle East, is recognized as the moment when a white thread cannot be told from a black. At that instant, under the ultraviolet light, the rocks blaze forth in their full luminous glory. The quarry heap or mine dump, if it is of the right kind, flashes forth with luminous color as the light is scanned over heaped rocks. There may be revealed the intense blue of scheelite, the red of calcite, or the green of silica or other minerals with traces of uranium. Often enough, brilliant combinations of various fluorescent minerals are to be seen, and the collector is hard pressed to select among the specimens at hand. In other cases, the fluorescence may be less brilliant, less varied, and even unrewarding, but this is the nature of the hunt, to be expected and accepted. Even at such times there are compensations, particularly in the mountains and deserts of the West where the chill wind which arises after darkness is bracing and the stars are unbelievably brilliant in a velvet black sky.

Night collecting thus means the chance to obtain good specimens. It means the thrill of discovery as darkness reveals the brilliant flood of color under the ultraviolet light. It also provides an often welcome solitude which is a contrast to the hectic workday environment.

It is thus interesting to review some of the experiences of collectors who, because of the diversity or uniqueness of collecting situations in which they found themselves, can cast light on the art of collecting and can project experiences we might duplicate, or at least wish to. Stories of two collectors – Tom Warren and Dick Bostwick – have been selected for this purpose. Their backgrounds are quite different, but their knowledge of fluorescent minerals is extensive, and their experiences are quite interesting and illuminating.

Tom Warren became interested in the fluorescence of minerals decades ago through his business, which is the manufacture of ultraviolet lights. He is the founder of Ultra-Violet Products, Inc., the largest and oldest maker of such lights. He has grown up with the fluorescent mineral hobby and has been a steady friend of the fluorescent collector as well as the prospector and geologist who use ultraviolet light. He recounts some of his experiences here.

SOME EARLY DAY COLLECTING

"I have been collecting fluorescent minerals since 1938 or 1939 and have seen the interest in fluorescent minerals grow from those early years. In the early 1930s, fluorescent collectors were using various makeshift lamps, such as the argon bulb, for display work. As for portable battery field lamps, there was little, except perhaps an occasional home-rigged iron spark or argon bulb working from an automobile battery.

"I became interested in fluorescent minerals as a result of my work. In the late 1920s, just out of school and recently married, I took a job working for a company which made neon tube lighting. The company decided to venture into the manufacture of ultraviolet lights to be used for health purposes, and I was put to work at that. But this was the depth of the depression. Sales were poor and the company could not sustain this work, so I was soon out of a job. Casting about for something to do, I decided to try to manufacture and sell ultraviolet lights on my own. With the help of a few backers, I built some lamps, sold some, and got through the first year.

"I was alert to any new application in which ultraviolet lights might be used, so that when a friend told me that miners might be interested in ultraviolet lights for assay or mineral exploration, I was interested. The first light which I made for this purpose was simply my general purpose ultraviolet light to which a filter was added. It soon became clear that a portable short wave battery operated field light was what the mining people needed, and so I concentrated on developing one. The first was a boxy affair, painted in black crinkle, with a leather handle on the back and a small window with a filter at the business end. It used a heavy duty 6-volt battery, and a vibrator and transformer to produce high voltage for the tube. This light was quite popular with the professionals. The main reason for their interest was scheelite. In the late thirties, the clouds of war were looming on the horizon. The production of tool steels and other strong steels was increasing. These required tungsten, and scheelite is the main source of this metal. Exploration was increasing, especially in the West, and particularly in California and Nevada where scheelite outcroppings are comparatively frequent. Scheelite is the most ordinary looking of minerals by daylight and it is very easily mistaken for so much worthless rock, but under the short wave light, things are entirely different. Scheelite lights up in a bright blue which is evident for some distance in the dark. So, portable short wave ultraviolet lights were indispensable to the miner and to the prospector searching for new scheelite deposits.

"It was about that time that I, too, caught the fever and began to go on night expeditions, looking for scheelite here and there in the deserts and mountains. Whether I should have been considered a prospector or a fluorescent mineral collector at that time would depend on whether or not I found any commercially valuable mineral riches. As I never did, I was clearly a fluorescent mineral collector, though I didn't recognize it at that time. However, others found values in plenty, using portable battery powered lights. By late 1941, the United States was at war. The conflict in the Far East cut this country off from the tungsten supplies of China, up to that time one of our primary sources of this valuable metal. Government priorities were set for the production of strategic minerals within this country and the scheelite boom was on. I worked overtime to produce ultraviolet lights required for this purpose, and these were indispensable in the location of new deposits and the efficient mining of the ore. With the aid of these lights, this country became self-sufficient in tungsten production.

"The fever for scheelite prospecting was high during these years. Every prospector and mining engineer was on the lookout for this essential mineral. Hundreds of small to large properties were discovered. Mills for processing the ores were set up in many locations, and the small mine operators trucked their ore to these mills to be concentrated to high grade scheelite.

"At that time, the U.S. Bureau of Mines also was carrying out exploration. While core drilling on a vanadium property in Idaho, they unexpectedly ran into rich sheelite a short distance from the surface. The mine company bulldozed off the surface layer and they scooped up the scheelite with a bulldozer. They worked at night and had a big, powerful, short wave ultraviolet light that we made for them on the front of the bulldozer. No lights of any other kind were allowed in the pits, and mining was done on the ore body by means of the fluorescence of the scheelite. I'm told it was the largest and most profitable scheelite mine in the United States during these war years. Several hundred scheelite prospects were discovered, largely with the aid of ultraviolet lights. Many were not extensive deposits. However, with the high price of scheelite, most paid back the cost of development.

"I continued my own prospecting as time would allow in those years, with no great commercial success, but with a good deal of enjoyment. Once I prospected in Death Valley, checking the hills and valleys, and of course, the washes. There was little to be seen but the expected fluorescence of borate minerals. On one of these trips to Death Valley, I met a young prospector who was investigating the Panamint Mountains bordering the valley. Each night he would head out in a different direction prospecting the slopes of the mountains for scheelite, but now his food had about run out. He was determined to give it one more try.

"Late that night, as he was returning to camp, his battery operated ultraviolet light suddenly revealed the fluorescence that he had been looking for. He secured a number of samples and the next morning showed his samples to the manager of a mine located nearby who, recognizing the value of the material, wanted to know where the samples had been found. It turned out that they had been found on the property of the mining company. The mine people had walked and driven over the area for years, and had not known that they owned a potentially valuable scheelite property until this prospector happened to discover it for them. He was later compensated in a satisfactory manner for the information and the discovery which he made.

"The end of the war in 1945 brought about a number of changes, including a major drop-off in the search for scheelite and in the need for the ultraviolet lights which I produced. My own increasing interest in fluorescent minerals, and the many people I had met during the war who were similarly interested, convinced me that there was a need for a good ultraviolet light for the fluorescent mineral collector. I set about to develop a modern line of such lights for that purpose. These included large lights which could be used for fluorescent mineral displays, hand held lights, and an improved portable. Most of these lights made use of the germicidal tubes which were becoming increasingly available as a low cost, dependable source of ultraviolet.

"The other result of the end of the war was that I had more time, and this meant more field trips. There were many, and there is no need to recount them all. I recall fondly an interesting visit to the benitoite mine in San Benito County, California. It is located on a ridge on the coastal range of California and is almost at the top of the ridge. It is a very interesting location, isolated from all habitation and extremely difficult to reach. This mine, the source of those beautiful gem crystals, has been open and closed, off and on, for as long as I can remember. The mine was abandoned at that time, and we were able to go over the dump rather thoroughly. However, what we wanted – those brilliant blue fluorescing benitoite crystals – was not abundant, and we found only sparse, broken specimens. The mine has since been opened and many very choice crystals have been recovered from it.

"However, a collector's fortunes shift, and the disappointing finds of one day are balanced by the successes of another, which happened when I made an extended visit to Arizona. I had been told that there were good scheelite crystals in quartz to be found somewhere in the Dragoon Mountains, east of Tucson. I suppose that I still had the scheelite fever to some degree, and I immediately became interested. Friends in a nearby mining camp gave me some directions to two potential locations. The first was a long vein of scheelite which surfaced for quite a distance. That night, a cold, rainy, very windy night, I was able to pick up quite a few scheelite specimens, but these were not the crystals that I was looking for. I remained the next wintry day, and that second night I found them. Where these crystals occurred was high on a steep ridge. It was difficult to work along the edge of the outcrop. I had to contend with clumps of brush, cactus plants, and a steep hillside. I had not checked this area in daytime, and it was slow, hard work on a dark night with occasional showers. The snow had melted, but it was cold. The crystals were at the bottom of the outcrop. These were pure, transparent quartz crystals. In the center, or the bottom, of the quartz crystals were scheelite crystals. I was unable to find any great quantity of these, but I found them extremely interesting, and while the night's collecting did not produce a great deal of material, I was satisfied. I experienced a thrill in getting the crystals because they were so unusual. The next morning was also unusual as I woke up with 4 inches of snow on my sleeping bag, which was a very strange experience for a southern Californian.

"I don't recommend this type of night exploration for everyone. It is certainly wiser to scout the land by day first, to understand better what the hazards might be and memorize landmarks so as not to get lost, and it is not wise to travel alone. As for getting lost, I have been lucky. I have a good sense of direction, and with the support of road maps or geological survey maps, I have been able to get to mineral locations without too much trouble. More important, I have found my way out of them and back to the main road without much trouble, even at night.

"Elsewhere in Arizona, at other times, I have found the most beautiful fluorescent willemite and calcite. At Ruby, near Nogales close to the Mexican border, I found the most brilliant red fluorescing calcite, brighter than the calcite from Franklin, New Jersey, matched in my estimate only by the fluorescence of calcite from Langban, Sweden. At Bisbee, at Superior, and in Moctezuma Canyon I found good fluorescing willemite and calcite, the better material being comparable to Franklin material. With some search, the material from Superior could be found with fluorescing fluorite also.

"The Moctezuma Canyon location was a bit unusual in that it did not appear to be a mine, but rather a cave created by water some eons in the past. In this cave, the ceiling and far walls are all willemite and calcite. In spots, the material was brilliant and beautiful – not all of it, but there were a few streaks that made it extremely interesting, even though the cave at times made me feel very uncomfortable since it appeared always to be ready to collapse. There is no doubt that the lure of the expected find, the anticipation of the beauty of the fluorescence as yet unseen, will draw one into caves or mines or other potentially precarious places where, with cooler heads, one would not go.

"The many red fluorescent calcite locations in Arizona can provide fine specimens for any collection, but one of the most interesting and beautiful red fluorescing calcites was to be found in the Cady Mountains near Ludlow, California. These were found in large masses almost as clear and transparent as glass. Within can be seen the fine lines of the natural cleavage planes of calcite. Etched along some of these lines, as if by an artist, are dark, almost black lines which form the shape of a large diamond figure within the calcite, and within this diamond another, and so on. Between these lines, the calcite fluoresced brilliant red in the shape of a large diamond. Outside the border, there was no fluorescence.

"On the first trip to find these diamonds, several other collectors and I were following directions given to us. We were able to drive over rough desert and reach a small hill about four miles from the road. We could get the car to the top of this little hill, but no further. The location that we wanted to reach was still over a mile distant. While the others stayed and formed a camp, I decided to try to reach the location before dark. I spent several hours checking one small canyon after another and finally found the narrow outcrop in the bottom of a ravine. Someone had discovered it earlier, of course, and had worked out all the easy-to-get diamond shapes. I stayed until after dark, and by means of the fluorescence, I found many specimens of these diamonds in the wash from the ravine. I left them all in a sack and proceeded to find my way back over the mile and more that I was from camp. The return was up a wash with large boulders, brush, and cactus. It must have taken me an hour or more to travel that mile. I found camp because those who stayed behind had started a small camp fire. The next morning we picked up the sack that I had filled. Two other trips to this same location were less productive. On the third trip, we found that someone had blasted the vein and it was completely obliterated.

"That sack, by the way, is my standard carrying equipment, nothing fancy. It is just an old potato sack. I generally travel light while hunting for fluorescents. In addition to this sack, or perhaps a few sacks, I may take a geologist's hammer. Chalk is also very handy. If you can't break a specimen out by night, or if you can't manage to carry it back, you will need to locate it when you return in the day. Since fluorescent mineral specimens are quite ordinary in daylight, it will be hard to tell which was the good find of the night before. The chalk can be used to mark the specimen when you leave it behind at night.

"The beginning of the 1950s saw a resurgence of interest in ultraviolet light. The government announced that premiums would be paid for the discovery of new sources of uranium needed to fuel reactors for the generation of nuclear energy, and of course, a great many uranium minerals fluoresce brilliantly. The scheelite prospector of a decade earlier was now replaced by the uranium prospector with his ultraviolet light and Geiger counter also.

"Certain uranium minerals make attractive additions to any mineral collection, and one can think of the beautiful green autunites of Washington State as an example. Other uranium minerals also provide attractive green fluorescent specimens, but equally attractive and of far more widespread occurrence are those minerals – calcites, aragonites, clear hyalite, and milky opal – which will fluoresce bright green when only a minute trace of uranium is present. Washes and coatings of hyalite can sometimes cover many feet of exposed rock, and the fluorescence can be seen for quite a distance. I can well remember one evening when I and some friends were in the Coso Mountains of California, looking for whatever could be found. We shortly came upon boulders some 20 and 30 feet in height, covered with masses of hyalite opal. Because of this large coverage, the bright green fluorescence could be seen as far as 50 or 60 feet away. The green fluorescence was quite bright. Some years later, my son and I were prospecting a site near Barstow, California, for fluorescent opalite. The opalite formed a cliff that rose up from the desert and covered a distance of approximately 100 feet. After dark we climbed over the piles of rock that had fallen and eroded from the cliffs. Some of it fluoresced a very brilliant green and other parts were a dull green. It was a cold, wintry night as we worked with hammers to break out the best fluorescent areas.

"The most attractive fluorescent opal comes from Virgin Valley, Nevada. It is highly prized around the world as a very interesting and bright fluorescent mineral. The fluorescent color is bright green due to traces of uranium. The size of the material which has been taken from this location is astonishing, some pieces of this milk white, glassy substance being several feet in size. The semi-opal comes in a vein form between layers of chalk. The region in which it is found is desolate and is mainly a cattle ranching area. In the same area, fire opal is also found. In some of the canyons near there we found quite a lot of petrified wood which was very fluorescent. I well remember the nights we spent in that area when the chill wind and bitter cold seemed to go right through the sleeping bags.

"I don't know why it was that so many of our night expeditions were in the wintertime. To those who may not be familiar with them, and who may picture the deserts of the Southwest as regions of perpetual heat, it may indeed be a surprise to hear that they may be quite cold at night even in the summer, and bitter cold on a winter's night. However, mineral hunting in the winter has one advantage, if only a psychological one. If one is looking at night during the summer, in many of the southern California or Arizona deserts there is always a possibility that the nice blue-green fluorescence on the ground may start to move. If it does, be sure you don't pick it up. Scorpions are very fluorescent. On two or three occasions I have almost picked them up; they moved just in time. Some wits refer to these blue fluorescent scorpions as "Arizona scheelite." As for rattlesnakes, I've been on several hundred night excursions and have never seen a live one. They are said to fluoresce also, but the dead ones I've checked never did.

"As a Californian, most of my collecting was in the western part of the United States, but I did get to the East now and then, and of course to Franklin, New Jersey. My first visit to Franklin was in 1946, and I made several visits at different times afterward. On one of these early visits, I was introduced to the mine superintendent and the resident geologist. I was allowed to go up to the picking table where the ore was carried and dumped on a moving belt before being dropped into the crusher. The man at the picking table removed all pieces of wood which might clog up the crushers. This also gave him the chance to pick out choice minerals, and he selected a number of minerals for me so that my pockets bulged when I climbed down the ladder to the ground. I never had this opportunity in later years when I had become more familiar with these minerals and might have wished to select my own.

"On these visits to Franklin, I met many of the miners and visited many of the basements filled with specimens. I had a battery operated light with me, and as I went through their basements, I was able to buy many of the fine specimens which they had accumulated. I bought large quantities of minerals from them during those years. The miners at that time when the mine was still operating were glad to find someone who would pay them for the minerals they had stored away. As a result, I have had for 30 years what I believe is the largest stock of rare Franklin minerals of any place in the world. At one time I estimated my total material from Franklin was between 50,000 and 60,000 pounds, all of it laboriously shipped to California.

"On trips to the East, I have also had the opportunity to collect fluorescent minerals in Canada. Canada, particularly in Ottawa and Quebec, is a treasure house of interesting minerals, including fluorescent minerals. I visited the site of the original wernerite discovery. A government geologist had mapped the area and had taken samples of different formations. He reported the discovery of the wernerite, and some was mined for a few years. The fluorescence of wernerite from this location was a deeper orange than that now coming from other locations, which tends toward yellow. On another property near Bancroft, a friend and I inspected a most dramatic outcrop. We found boulders that must have weighed a ton which brilliantly fluoresced blue and yellow – diopside and apatite. Not all of it fluoresced, and there was a great variation so that selectivity was important. I couldn't help thinking that here was a beautiful mass of fluorescence a hundred miles from anywhere. How did anyone ever find the location? It had no real commercial importance, but some prospector had used an ultraviolet light and found it. If it was near a metropolitan area, thousands of people would want to see it or want to collect here.

"Whenever I looked over a fluorescent area, I often wondered about the person who had first discovered it. What was he looking for? How did he happen to be in a particular location? I knew that without an ultraviolet light they could never have found the fluorescence. I also knew that hundreds of locations must have been checked in order to find one of importance. It made me realize that there are many more discoveries yet to be made. It's a thrill to find something new – to be the first to make a discovery. You can never tell what you may be walking over. Without the advantage of an ultraviolet light, you may be walking over a treasure of fluorescent beauty that is under your feet and you would never know it."

Tom Warren could have related much more, for his experiences in collecting fluorescents span more than five decades. The many diverse and varied minerals he has collected and the territory he has covered, provide some idea of the vast scope for collecting which is available in this country.

However, as his experiences have been wide ranging, geographically speaking, the following section describes collecting in the narrowest of confines – the stopes and drifts of the famous Sterling mine in Ogdensburg, New Jersey. This mine is second only to the long closed mine nearby at Franklin in terms of the rich variety and beauty of the fluorescent minerals found. Many a collector dreamed of the fluorescent riches which might be had if only it were possible to spend an hour or two in a mine, but particularly this mine. Here a magnificent and rare fluorescent specimen might perhaps be found before it is sent to the crusher, before the miner and then the dealer claim possession, before it is high-graded unseen into the hands of well-connected collectors. Dick Bostwick fulfilled that dream. A New Jersey native, he graduated from Yale and then worked at a succession of jobs before returning to one of his earlier interests, mineralogy, taking a job in the mineral department of Ultra-Violet Products. After some time, personal matters required a return to the East. By now, fluorescent minerals had captured his imagination. Hearing that the Sterling mine was hiring, he applied for a job and soon found himself in the dark of the mine. He spent years there as a miner. He finally left this demanding occupation no worse for the experience, but with fewer illusions about the work of the miner or the opportunities for collecting. Dick Bostwick has remained a collector of Franklin and Sterling fluorescent minerals, and is recognized as a leading authority on this subject. His story of those years in the mine is illuminating.

IN THE STERLING MINE

"Every fluorescent mineral collector knows of the minerals from Franklin, New Jersey. Their uniqueness, fluorescent brilliance, and rarity have made them showpieces in any comprehensive display. Most of these collectors know as well that the Franklin mine closed in 1954, its rich zinc ores exhausted and its miles of drifts abandoned to the rising groundwater. Since that year, esperite, hardystonite, clinohedrite, and other distinctive minerals from Franklin have been attainable only from aboveground sources: miners, collectors, dealers, and – with luck and perseverance – the dumps. Supplies dwindle yearly, prices rise, and among Franklin enthusiasts there is much nostalgia for the Good Old Days, when the Parker dump regularly yielded esperite and a few dollars or a bottle of whiskey would liberate superb specimens, fresh from the finder's lunch pail.

"Enveloped in the haze of hindsight, we tend to ignore the existence and continued operation of the Sterling mine, located in the east flank of Sterling Hill, across the Wallkill River from the town of Ogdensburg. Here franklinite, willemite, and zincite are still mined, milled, and shipped away for refining; here the miners drill and blast, and bring out heavier lunch pails at shift's end. Most collections have specimens labeled "Franklin" which came from a different shaft over two miles distant, carried out by miners who never worked at Franklin.

"Among fluorescent mineral collectors in particular, Franklin's reputation has always completely eclipsed that of Sterling Hill. Recently the Sterling mine has become notorious among mineralogists and knowledgeable collectors as a source of rare mineral species. Most of these are found in small amounts, often as microcrystals; more than a few are unique, new to science. Specimens from that mine are now examined most carefully, and this close scrutiny has also led to the identification of several fluorescent minerals never found at Franklin.

"Along with a desire to walk the streets and dumps of pre-1954 Franklin, most fluorescent mineral collectors share a related dream, that they could go underground and confront their favorite minerals in person. Armed with hammer, chisel, and ultraviolet light, what collector would not want to track the wily willemite to its subterranean lair? I satisfied this longing for a period of over three years by working underground at the Sterling mine, first as a helper and later as a drill runner. Until then I had regarded miners with envy and a little awe; I scratched around on the surface to find what they had thrown away, or I slaved in my own salt mine to earn money for specimens which cost them nothing. Each piece in my own collection had been acquired with effort and sacrifice, while those miners were undoubtedly trampling crystals into the mud or shattering them with dynamite. These thoughts spurred my interest; even though I had spent ten weeks in the same mine as a summer job years before, I was not discouraged nearly enough. To my imagination, the miners still labored like gnomes in dark chambers which were solid fluorescent masterpieces from top to bottom. What an adventure, I told myself, to *be* there, an apostle bringing the gospel of glow into the depth of the earth.

"Now and then in collectors' faces I encounter the same desire, the same fantasies, envy and awe alike. However, working underground is not an endless field trip, and in many respects working the surface with a silver pick is better than going down and getting dirty. What follows is part history, part opinion, and hopefully a partial antidote to that unsettling disorder, mine fever.

"The Sterling mine is unusual in that its miners are allowed to take specimens home. Most mines forbid it outright and fire their men after the first or second offense – at least, so I have heard. The following rules were posted in the adit of the Sterling mine in February of 1978:

1. Small hand specimens, no larger than the size of a closed fist, that will fit in a lunch bucket, may be taken by an employee for his private collection.
2. Specimens may be taken only from the employee's immediate working area.
3. There is to be NO traveling through working places searching for specimens. To do so is an exceedingly dangerous practice which could result in severe injury to anyone not familiar with the working place he has entered. Anyone found violating this rule will be subject to discipline.

"All things considered, this is a remarkably enlightened policy. The minerals of the Sterling mine are in many cases extraordinary or unique, and most of them would never have seen the light of day except for the miners. Were collecting prohibited, this would greatly restrict future scientific knowledge of the deposit, as well as frustrating hundreds of active collectors. In practice, it is questionable whether specimen removal could be banned; most Sterling miners consider it one of their perquisites. Some of the more active rock hunters sell their booty for a tidy second income, while others stash it away with an eye toward retirement. More than one zinc miner has depended on rocks for beer money or for putting a child through college. Still others collect for themselves, because minerals fascinate them. For a few miners, the obsession is as strong as it might be for any collector, and a major reason to stay underground.

"Statistically, mining is one of the least safe occupations. The Sterling mine has a fairly good score on that account, but the New Jersey Zinc Company's concern for miners collecting outside of their working places is not unjustified. Even though efforts are made to keep barriers between miners and the off-limits areas, the inquisitive drive can be very strong. Also, the Sterling mine is old: some of the open pit workings date from before the revolution, and extensive underground development dates from the time of the First World War. Many sections of ore have been mined out and filled, but the access drifts are still there, tunnels to nowhere, dark and silent as Tut's tomb. There are shafts and even entire levels that have been abandoned and can still be visited. At the openings of dead-end drifts sit obsolete ore cars, patiently rusting back to ore themselves. Crumpling concrete barriers proclaim themselves the home of ghosts: Luke the Spook, Bicycle Pete. The connoisseur of relics is likely to trip over hand-sharpened drill bits, shaft and level signs long out of use, and other remains of Sterling's past. There are also rotten timbers festooned with impossible fungi, as well as loose slabs, flooded drifts, foul air, and strange smells.

"All that really can be said is that the company did not pay us to sightsee and that there were objective perils in being away from one's working place at all. My own investigations were generally confined to the area I had been assigned and to times when there was nothing else to do. Machines were always breaking, supplies ran out or didn't arrive in time, one's stope or pillar had to ventilate after a blast, and there was always lunch hour. As with every job, a few liberties were permitted on a "see no evil" basis. One could be reasonable and circumspect about one's wandering and collecting, or one could be a damn fool and spoil it for everyone.

"Given these conditions, how many miners actually collected anything? Most of them would, if an obviously valuable crystal was sitting on top of the muck pile. A few made a habit of collecting, taking home anything that might possibly prove worthwhile; I knew of only one who collected systematically and intelligently. Many, perhaps the majority, were either indifferent or actively hostile toward "specimens." After all, a miner's job was (and is) to get the ore out, not admire it. Interrupting the mining cycle to save a rock from its rightful destiny in the crusher appeared to be softheaded. Nor was it properly macho to display concern for the safety of delicate crystals in an atmosphere where miners were frequently hurt and didn't whimper about it. The first decent specimen I found was taken from me and smashed by a miner who wanted to show me how he felt about rocks, not to mention those who collected them.

"In his indifference toward, or contempt for, the rare and fluorescent minerals of the Sterling mine, that miner is not alone. Only the enthusiast would find this attitude unnatural. Most of the local rarities are either massive and ugly, or tiny crystals which are aesthetically pleasing only under high magnification. Show such things to the average mineral collector, who is biased in favor of the attractive specimens displayed at museums and mineral shows, and unless he is both knowledgeable and tolerant he will be indifferent or actively contemptuous.

"In the mine, a miner who picks up obvious crystals will probably miss very small ones and ignore fluorescent material altogether. It is a burden to buy an ultraviolet light and take it underground; it leaves little room in the lunch pail for rocks. In the latter days of the Franklin mine, matters differed: more than a few miners owned ultraviolet lights and collected fluorescent minerals with great enthusiasm. Esperite and others were bright and beautiful, easy to find, and thoroughly enjoyable without any knowledge of minerals. The light of the period was the M-12 Mineralight, a black box 3 X 3 X 10 with a filter less than 2 inches square. I have heard this ingenious device cursed because many of the miners carrying it ignored the rarer nonfluorescent minerals. My opinion is that without that light and the beauty of the specimens it revealed, most of those miners would not have collected much of anything.

"The portable ultraviolet light seen underground at the Sterling mine in the M-12s descendant, the M-15. This rechargeable long wave–short wave lamp fits a largish pocket, and is carried by the geologists and engineers of the New Jersey Zinc Company. It has a practical use: determining the richness and limits of the ore. Willemite, one of three major ore minerals at the Sterling mine, fluoresces a characteristic bright green. Of the other two, franklinite is nonfluorescent and zincite almost never, but willemite generally occurs with both. To the "Company," if the rock fluoresces green, it is rich enough to mine. This on-the-spot determination is much simpler all around than sending samples to Palmerton, Pennsylvania, for analysis, and in fact it works very well most of the time. Allegedly a high grade horizon of very rich zincite-franklinite ore was temporarily ignored because there was no willemite present, and no green fluorescence, but such oversights are rare.

"If you took such a light on a field trip underground, what would you see? Very little, unless you are lucky and there is freshly broken ore in certain working places. The main shaft and the working drifts throughout the mine are in the Franklin marble, the same rock obtained from local quarries. Only within a few feet of the zinc ore is there likely to be enough manganese to cause the characteristic red fluorescence of calcite. In the drifts you might find fluorescent tremolite and norbergite, but these are scarce and hardly worth the trouble, if one considers that the same things are comparatively abundant in the quarries aboveground. To make things worse, the drift surfaces are covered with a fine powder settled out from nearby blasting. As willemite is usually present, most of the drift walls will fluoresce a faint, eerie green regardless of what minerals lie beneath. Where drilling is recent, your lamp will reveal bright blue spots, splotches, and smears all over. This is not scheelite, but the heavy oil used to lubricate rock drills. It is tenacious and persistent, highly resistant to soaps and scrubbing, and more than likely to coat the fine specimen you have just picked up.

"Dust and drill oil notwithstanding, suppose you luck out and show up in a working place where the ore is newly blasted and just washed down? The chances are you won't find much – except ore. There will be an abundance of willemite, and it will fluoresce the customary green, but the odds are that the calcite with it will be dead under the lamp. Most Sterling mine calcite has too little manganese to fluoresce well, or too much. Now and then, in certain parts of the ore, the manganese percentage is just right. Ore specimens from those areas will have an intense red and green fluorescence matched only by the choicest Franklin pieces. Incidentally, the abundance of this combination has somewhat obscured the fact that the finest Sterling mine willemite-and-calcite ranks aesthetically with the very finest of the world's fluorescent specimens. Even if you're in the midst of a mound of it, however, there is no guarantee you'll find one of these matchless showpieces. Specimen status is determined by such aspects as pattern and color balance, and nothing about the ore's structure or the physics of blasting assures choice items. Usually these seem to be attached to the walls and overhead, or incorporated in fat chunks. The mine management has thoughtfully provided specimen trimmers in the form of sledgehammers and axes, but the results obtained with these are often less than optimum. Try it on your own specimens sometime.

"So if you are in the right place at the right time, and if you brought your light, you may just find yourself in Aladdin's cave for a few minutes. As mines are absolutely dark, a small light will set fire to a moderately large patch of ore, or evoke a phantom glow from a much larger area. During my term underground there were only three or four working places which yielded high quality willemite-and-calcite, however, and their existence was no guarantee that you as a miner would ever collect there. The mine was divided into three administrative units: Upper Section, Middle Section, and the Lower Section/North Ore Body. Each section had its own bosses and crews. Although some miners would transfer from section to section, and others had jobs that took them all over the mine, many more would stay in one section year after year. Working overtime on Saturday was the only sure way to get out of your section, and under the conditions then prevailing, a person could work in the mine for 20 years and not see all of it. Furthermore, the mining methods at Sterling Hill dictated that collecting in any one place was occasional at best. The mining cycle included drilling, firing, roof bolting and timbering to secure the ground, "mucking out" (removing the ore), fill preparation, and the actual filling. The best time to collect was during the roof bolting and mucking out phases, when an abundance of broken ore lay there to be hosed off and scrutinized. During drilling and filling, there was little ore that wasn't attached to the earth or exceedingly dirty. Furthermore, although the overall character of the ore in a stope or pillar was generally consistent from cut to cut, collectible specimens might be abundant one cycle and absent the next. There is nothing quite like the thrill of being drill-runner-for-a-day in a legendary working place and discovering that the cupboard is bare or buried.

"Willemite was not the only fluorescent ore mineral, even though it was the only one the mine management used as an ore indicator. Sphalerite, the major ore of zinc worldwide, was quite rare at Franklin. In the Sterling mine it is widespread in small quantities, occasionally making up 1 percent or more of the ore, and for variety of form and fluorescent response it is remarkable. Sphalerite found elsewhere is generally black or brown and loaded with iron, which is a very effective fluorescent quencher. Most Sterling mine sphalerite is iron free, the variety called cleiophane; its color in daylight is pale yellow or gray for the most part, but it can be buff, dull orange, pale green, maroon, or a most unusual light blue. Fluorescence under long wave ultraviolet is likely to be orange, but varies to the striking orange-yellow of "golden sphalerite" and the bright blue to which local collectors attach the title of cleiophane, rightly or wrongly. From a mineralogist's point of view, the orange fluorescence may be activated by manganese, and the blue by silver, while both varieties are free of iron and therefore are cleiophane.

"The first working place I took over as a drill runner had sphalerite in occasional abundance, as pale yellow glittering masses. This working place was a square set pillar, where the ore is removed in units about 5 feet 8 inches square and 8 feet high, and the ground is supported by a rectangular network of interlocking timbers. Blasts were small, mining was slow, and much of the mucking out was done by hand; consequently there was ample opportunity to collect. At one time, the plank floor on one level of the pillar was covered with hundreds of pounds of sphalerite. Every visitor helped himself, and for weeks I lugged out a full lunch pail every night. I blush to admit salvaging many pounds of this material and do not care to speculate about how many tons have been removed in the history of the mine. This example of Nature scattering her treasures with a profligate hand contrasts nicely with the situation in a neighboring pillar, where the prize was "golden sphalerite" with its odd orange-buff color and remarkably bright fluorescence, golden orange laced with blue. Here hours of vigilance and knuckle scraping were rewarded with perhaps 50 pounds of miscellaneous glop, only a few pieces of which were larger than fist size. The word "glop" is used deliberately; the area had undergone extensive water etching, and most of the sphalerite had been reduced to crumbs and paste. Collecting this could be very disheartening, especially when an hour's careful work yielded only five 2-inch mudballs and 10 pounds of fluorescent sludge.

"Zincite is zinc oxide, red in color, ranging from almost black to a bright blood red. It is the third major ore of zinc at Franklin and Sterling Hill after willemite, which fluoresces conspicuously, and franklinite, which is black under daylight and ultraviolet light alike. Until fairly recently, zincite was not thought to fluoresce at all. In the early 1970s, veinlets of a yellow mineral in zincite rich ore were discovered in 980 stope above 500 level. Not only was this "new" mineral attractive, it fluoresced yellow under long wave ultraviolet light. Arsenic was diagnosed, and the unknown enjoyed a brief vogue as adamite, judged vaguely similar to the fluorescent crusts of adamite from Mapimi, Durango, Mexico. Under shrewder eyes, the Sterling Hill fluorescent adamite became zincite, a very pure zincite without enough elemental garbage present to quench its fluorescence. Several years later, when 980 stope had been mined out and the same section of ore was being mined by the 800 stope three levels up, fluorescent zincite was still turning up. On my one Saturday visit to the location, I was shown where copious quantities of this zincite had been just the shift before. Although common associates are fluorescent willemite and calcite of the conventional sort, these yellow zincite lenses may also be found with blue fluorescing hydrozincite, sphalerite, and/or fluorite, as well as a very unusual calcite coating which fluoresces orange. This material is overlooked by fluorescent mineral collectors in spite of its potential for containing seven different fluorescent minerals and five fluorescent colors in one specimen. I haven't seen one, but perhaps the Gentle Reader will.

"Perhaps more frustrating than missing good stuff because one didn't look for it is looking for something diligently and not finding it. My first two years and five months underground were spent in the North Ore Body of the Sterling mine, and of all the sections this was most nearly free of fluorescent minerals. Willemite was abundant, but the calcite with it was dead under the ultraviolet light; the occasional patch of fluorescent calcite was usually outside the ore. My excitement over the sphalerite already mentioned was due to starvation. Otherwise the only fluorescent thrills lay in cruising the drifts, looking for secondary minerals on the walls and in cracks. Now and then small streaks of hydrozincite appeared, pleasing when on a matrix of fluorescent calcite; in those 29 months I was able to collect perhaps one pound of this. As an added attraction, there was the widespread presence of what a fellow miner christened the Green Slime. Damp areas in some of the drifts were coated with it; occasionally it brightened to a canary yellow, but on the whole looked quite unhealthy. Finally in desperation I took home several lunch pails full of the material. After all, it did fluoresce a bright uranium green. No one is quite sure what the Green Slime is, other than an impure, highly hydrated dolomite. No one has felt like finding out exactly what is causing the fluorescence. My sole consolation is that after several years of drying out, the mineral or whatever it is no longer looks slimy. Now it most closely resembles a film of spilled lemon yogurt left in the sun.

"When my North Ore Body pillar ran out of ore, I was sent several hundred feet upstairs to the Lower Section. At first my helper and I were assigned to "drift rehabilitation," barring down loose rock and securing shaky ground with roof bolts. Here the Green Slime was mercifully absent, but there were many cracks; near the ore you could find hydrozincite. At one turn of a drift on the 1750 level, sphalerite rich ore with numerous slip surfaces was exposed. Here groundwater had been at work for a long time, and each new slab pried from the drift wall was covered with hydrozincite, richly so. My frenzied efforts yielded one large specimen, one medium, and several lunch pails full of minis. I flattered myself that I had gotten all that would come easily, but failed to reckon with the determination and ingenuity of the zinc miner when faced with an extremely obvious specimen source. In the following months the appearance of the drift kept changing: it grew wider and wider. Pry bars and sledges littered the drift floor, along with slabs and chunks of rock. Hundreds of pounds must have been removed before the novelty wore off and the hydrozincite ran out.

"The Green Slime turned up once more, with surprising results. I had been working with the raise bore drill, a large growling beast of a machine which drilled holes 4 feet in diameter. Something went wrong, so my helper and I took a walk to the north end of the level. There we noticed a pale yellow goo apparently oozing from a drill hole and trickling down the drift wall. I had my ultraviolet portable along that day, so we dutifully checked the stuff; it fluoresced the same uranium green as the North Ore Body material. By then I was heartily tired of such things, but we helped ourselves out of desperation. The miners who saw it commented that it looked too much like a bodily excretion to be a mineral. Eventually one of them took a piece home. He sold it to a mineral dealer, and the dealer sent it to the Smithsonian; this august institution proclaimed it a mineral new to the area, and a rare one at that – monohydrocalcite. The tangled history of its discovery is typical of many Sterling mine mineral species. I noticed it first, or I may have, or at least I don't remember anyone else telling me about it first. My helper Ron is the one who insisted on taking some home. As a result of his interest, miner No. 3 (Big Jim) appropriated the piece which tickled the dealer's interest and fogged the film in the Smithsonian's camera in just the right places. From there the mineral leapt into print, on p. 163 of the May–June 1979 *Mineralogical Record.* If any step in this sequence had been aborted, the mineral probably would have remained uncharacterized and unsung; certainly in my collection it would have been another nameless piece of junk. For once there is a happy ending, however. The mineral has a proper name and the dignity of being in print, no matter if it resembles a blob of glop.

"The history of North Ore Body dypingite is somewhat similar. My first regular assignment there was in No. 3 pillar below 2250 level, as an unusually green and helpless helper. There I could not help noticing a mineral which was widespread toward the footwall of the ore. Where the ore had been cracked and weathered, the surfaces were black with manganese stain. Against this background were abundant tiny hemispheres of a pale blue mineral, often in curious patterns suggesting an Eskimo housing development. I might have taken out more than a few pieces if I had not brought one to a knowledgeable collector for identification. He suspected calcite and dribbled some dilute HCl on the ranks of mysterious igloos. They fizzed and vanished: Q.E.D. By this time the pale blue had bleached to white, and I would have discarded the pieces except that they fluoresced a faint blue. In my collection were perhaps 20 varieties of fluorescent calcite, so the faded newcomer was shoved in with them. Three years later, a piece went to the Smithsonian as part of a small North Ore Body collection and was identified as the hydrated magnesium carbonate, dypingite. It was previously unknown in the area; once recognized, it started popping up in many places underground and is now not as rare as some collectors would like to believe.

"Chance plays a strong role in these affairs. Not a few of the rarest Sterling Hill minerals were found once only, in small quantities. Others were locally abundant, but only a few pieces were saved. When so many rare or unique minerals are either very small or very ugly, it is difficult to save them, especially if they haven't been identified yet. It is fruitless to speculate how many tons of extraordinary Franklin and Sterling Hill minerals have been turned into paint pigment and carburetors. We can only be thankful for what survives. If it were not for the dirty and destructive work the miners do, nothing would be known of these minerals at all.

"Speaking of miners, would you want to be one, assuming that you might be able to collect minerals on the job? A few sturdy souls might take up the challenge, but in fact no one who is primarily a mineral collector lasts long at the Sterling mine. Those who both mine and collect enjoy mining as a way of making a living. You go underground every day and count on having your job to do; you can't count on having a lunch pail jammed full of specimens at shift's end. Mining is dangerous, suprisingly complicated, filthy, often frustrating, and always tiring. If your head is throbbing from dynamite fumes, or if your body shakes from having hung on a rock drill all day, each little scrap of rock loses some of its charm. It is worth remarking that underground, rock is The Enemy. You are struggling endlessly to keep it under control, either moving it from place to place or keeping it from falling on you. Every miner knows that a rock can kill him without warning. It is treacherous even when you are sure of yourself: a fat chunk of ore can roll over and crunch your ankle, and calcite cleavages bite like razors. After days or years of fighting The Enemy you may not regard it in a friendly light. Furthermore, good mineral finds aren't common. In my experience, 6 to 12 weeks might elapse between one bonanza and the next. In the interim you could easily go stir crazy, staggering up and down the drifts like Ahab searching for the white whale. I spent a good deal of time kicking rocks in the hope that a mineralogical spring in the wilderness would flow for me.

"It didn't help that many Sterling mine minerals are hard to identify. In practical terms this means that you can't tell what they are by looking at them. Common minerals often masquerade as rarities, and vice versa. My strategy, when I thought of it, was to take home anything that looked "different," while staying attuned to slight variations in appearance, occurrence, association, and so on. This can be difficult when all the rocks around you are covered with rock dust and drill oil, and you are hurrying to wire up a blasting round before the end of the shift. I quiver with anguish to remember how many nameless mineralogical horrors I hauled home in good faith, only to discover after washing that I had yet another variety of serpentine or the world's ugliest blob of willemite.

"Curiously enough, there were fluorescent displays underground that could be enjoyed without benefit of an ultraviolet lamp. One of these was fox fire, the spectral glow given off by fungi on decaying timber. In the damp darkness of the mine, pine planks 3 inches thick could become feather light and crumbly in a year. Not bright, this bioluminescence could be seen when your cap lamp was out and your eyes adjusted to absolute blackness – as in the case of waking up from a nap. Drilling in the dark, if you chose to try, could generate a flickering light around the drill bit; much willemite is slightly triboluminescent, as is sphalerite. Perhaps more remarkable is the phenomenally sensitive fluorite found in and near Sterling's "black ore," which is a drab mixture of black franklinite, dark gray or black willemite, and dull, non-fluorescent calcite. This fluorite is found in sherry colored grains and masses, both within the ore and in the barren calcite nearby. Pass the beam from your cap lamp over the fluorite, switch off the lamp, and a pale green luminescence will be seen. This response to incandescent light is similar to the fluorescence of the mineral under short and long wave ultraviolet light; unfortunately, the whole effect is highly perishable. Exposure to daylight for a few hours will destroy the mineral's light-sensitive phosphorescence, and eventually it will only fluoresce a dull blue under an ultraviolet lamp, like many other fluorites. One miner I knew lost a fair amount of money by betting that his special rock would glow in the dark, like a pair of plastic Halloween fangs. "No kidding, this stuff glows after you hold it next to a light bulb, honest, wanna bet?" He had left it in daylight too long and after that wasn't quite as interested in minerals. Friends became skeptical of his mine stories; his reputation was damaged – many are the hazards of collecting. Other efforts of mine to bring enlightenment to dark places met with equal success. After watching me scramble for specimens month upon month, my helper finally took home one piece of ore, sprayed it with glossy transparent plastic, and displayed it next to his TV. Another miner let himself be coaxed into viewing my fluorescent display. For half an hour I waved lamps and rocks around, dancing with enthusiasm, without any response worth mentioning. I raved about the colors, their brilliance, intensity, and contrasts, to a poker face. Finally I forced him to react: "Look, Tricky, I hate to tell ya this, but, like, I'm color-blind, ya know?"

"My thrill of thrills underground came while collecting fluorescent wollastonite. The location was in the wrong place, to start with. It was on the other side of the level, protected by a stout brattice (a fence-like barrier intended to keep inquisitive miners out of dangerous areas). The drift floor was littered with slabs, and more were peeling away from the walls. Several roof bolts had become "chandeliers," dangling steel rods now serving only to suspend a large chunk of ore over your head. Oddly enough, it was safer than it looked, as the drift had gradually acquired a rounded-arch cross section, structurally very sound. Because the drift wall was clean and freshly exposed, the wollastonite could be traced for many feet. When eyeballed, the drift was a tunnel in barren white marble; dosed with a portable ultraviolet light, it was a tapestry of orange and golden yellow spots, bounded on one side by brightly red fluorescing calcite. In one narrow band this calcite overlapped the orange fluorescing wollastonite, for a striking two-color effect. Inches away the calcite ceased to fluoresce and the wollastonite turned yellow; another few feet, and the yellow grew paler and dimmer, to be traced for another 10 feet at least. An impressive exposure, to be sure, but others had been there before me and work was necessary. Half the fun was getting there, carrying a pry bar, chisels, ax, tote bag, and portable light perhaps a thousand feet through muddy, uneven drifts. Having arrived, one had to break through the brattice without inflicting visible damage. After this came the collecting, such as it was, with the usual clunks, clangs, and foul oaths as one's knuckles ran red and the specimens fell into the oily, omnipresent mud. Now it was time for the awkward crawl back through the brattice, the virtuous act of nailing it shut, and the long stagger back to one's working place. All this, of course, was within the compass of the half hour allowed for lunch. The thrill alluded to came from nearly being caught in the act of collecting. I heard splashy bootsteps, thought it was my helper, and decided to hide from him. As I was on the verge of howling like a gorilla, I recognized – the shift boss! There I crouched, lamp off, trying to look like a rock and rehearsing excuses: "Nobody here but us specimens, boss."

"With or without this brand of excitement, mining at Sterling Hill is not a smorgasbord of mineralogical delights. Worse, most of the material brought out is what one prominent mineral dealer describes invariably as "poop." The average miner's collection consists of a handful of superb pieces, a peach basket of intriguing ones, and half a ton of very miscellaneous ore samples. Reflecting on this should help you appreciate your own collection. Most of the specimens there will have already passed through several levels of selection. With your choicest minerals you are standing at the tip of an enormous pyramid, its base far underground, with those who built it, like Pharaoh's numberless slaves, invisible and silent.

"The Sterling mine has some life remaining, although no one knows exactly how much. Before the year 2000, sooner or later, all the minerals that can be mined at a profit will be gone, and Sterling Hill will join Franklin on the list of extinct localities. The golden age of collecting there is now. For the ambitious rock hound, I have two bits of advice: get what you can while it's there, and get it in some other way than going underground."

BUILDING A COLLECTION

Most fluorescent collectors start as collectors of nonfluorescent minerals, and they usually have a basic understanding of minerals and of how a collection can be built and displayed to advantage. Fluorescent collecting follows the same principles with a few peculiarities of its own. The problem which the beginning collector will face is how and where to obtain fluorescent specimens. As in forming other mineral collections, there are several means. The mineral dealer is a good source for fluorescent specimens, and the names of dealers nationwide can be found in the various lapidary and mineral journals. Dealers at well-known fluorescent localities usually feature fluorescents of that locality and are among the best sources of the best and most interesting specimens.

Other dealers are often not particularly aware of the fluorescent minerals they may have. They may offer a few Franklin fluorescent specimens, or perhaps a wernerite from Canada, but be completely unaware of other possible fluorescents in their stock. Therefore, when visiting dealers, it is useful to bring an ultraviolet light and obtain permission to examine the material. As discussed earlier, a dual band light is particularly useful for this purpose as it permits either short or long wave fluorescent minerals to be detected with a minimal survey. If there is too much light in the area, specimens can be examined under a table, an opaque cloth, or a jacket. Some visible light will leak in, and this will distort the true fluorescent color and intensity, but the presence of fluorescence can usually be detected. While fluorescent minerals are infrequent, usually a few surprises may be found in any dealer's stock. Generally, a few percent of dealer specimens can be expected to fluoresce.

Good material can also be obtained in trade with other fluorescent mineral collectors. Others who collect fluorescents should be identified at mineral club meetings, and an eye should be kept open at shows for fluorescent collectors who may sometimes be found near the stands of dealers who sell fluorescent minerals or equipment. Mineral club swaps are also rewarding. The Fluorescent Mineral Society, with membership in many countries across the world, counts among its members a number of collectors with trade material. Since Fluorescent Mineral Society membership is far spread, trade by mail is the common thing. Specimens traded sight unseen call for particular care in selecting and describing the material offered. Such material should be accurately described in terms of specimen size, size of fluorescing area, color, brightness, and other factors which will be needed in order to form a good picture of the specimen offered. Only good material should be offered, material which you would not hesitate to have in your collection. Poor or disappointing material dries up trade, a consequence which hurts all collectors.

A primary means of obtaining fluorescent minerals is by collecting by day or night. It is possible to use a portable ultraviolet light in the daytime on a promising dump. For this purpose again, a large opaque cloth is needed, and the material will have to be examined under this cloth. Many fine fluorescent specimens were found at the dumps at Franklin, New Jersey, with the use of a blanket and a portable battery powered ultraviolet light. This approach is admittedly cumbersome when a great deal of territory is to be covered. It is also difficult to carry out under a bright sun. This problem is not so much one of shutting out sunlight. Rather, it is the desensitization of the eye caused by the continuous intense glare, particularly in quarries, which bleaches the visual pigment in the retina. This problem is usually less serious on an overcast day.

Night collecting by use of a portable ultraviolet light can be the most rewarding and satisfying way to obtain good fluorescent minerals, but this calls for good preparation, since collecting at night can be hazardous. All of the hazards of quarries, mine dumps, and other collecting sites are magnified in the dark. It is easy to misstep, to miss a footing on a rock pile, or to forget where the edge of the quarry bench is located. Further, another hazard exists, particularly in the West. Here, many exploratory pits, air shafts, or working shafts are often open and uncovered, without fences or warning signs. These can be death traps if stumbled into in the dark. As a consequence, special precautions should be taken for any night collecting. The following three are the most important:

1. Avoid locations which are hazardous by day since they will be much more hazardous at night.
2. Keep to a location which has been thoroughly scouted by daylight, and commit its main features and landmarks to memory. Keep a constant back check on the location of your car, or a road leading to your car or camp. These references are easily lost at night as one moves about, eye to the ground; once they are lost, disorientation is possible.
3. Don't explore alone. Two, or preferably more, people should make up the party.

In addition to the usual mineral collecting equipment, there will be needed a good single or dual band portable light and a good flashlight, tied with a string or rope which can be looped around the neck or through a belt. In addition, a role of masking tape is particularly handy during a night search. Strips of tape can be attached here and there along the line of exploration to mark the trail back. Such strips are surprisingly visible by moonlight or even starlight, and a flashlight may not be needed to detect these biodegradable trail markers. Also, this tape can be used to seal the paper wrapping which is placed around specimens found. For wrapping, plastic bags or brown paper bags should be used since newspaper contains fluorescent dyes which have a way of transferring to sharp edges of a specimen when it is buffeted in the carrying bag. If possible, all tools should be marked with a fluorescent tape so that misplaced tools can be easily spotted in the dark with the ultraviolet light.

A word of caution is needed concerning trespassing on private land. Most land in the United States is privately owned. Active quarries and mines are usually fenced and patrolled, and permission to enter will need to be solicited if one is to avoid trouble. The same can be said of dormant and abandoned mine and quarry property, though in many cases these are not under guard or fenced, and as a practical matter perhaps no one cares about a trespass which does no harm to the property.

However, all too frequently harm is done. Unthinking visitors to abandoned mines and quarries have damaged property, littered, left gates open, and in other ways not shown good sense and good manners. That this is not a recent problem is indicated in an article on this subject in the April 1939 issue of *Mineralogist*. The article included a photograph of a crudely made hand lettered sign nailed to a chained gate saying:

> NOTIS! tresspassers will B percecuted to th
> full extent of 2 mungrel dogs which neve was
> over sochible to strangers & 1 dubble brl shot
> gun which aint loded with sofa pillers.
>
> Dam if I aint gitten tired of this hell raisin on
> my place.

This probably expresses the sentiments of many owners of property upon which mines and mineral dumps are located. Irresponsible behavior on the part of those who have entered is the cause of the problem. To be welcomed, and to allow those who come later to be welcomed, sensible behavior while on the property of others is mandatory.

With the foregoing precautions in mind, it is possible to have an exciting and rewarding night exploration, and to add interesting and attractive specimens to the collection.

PREPARATION AND DISPLAY

Once the material is home, is should be examined to identify the constituent minerals, to confirm fluorescence, and to determine the fluorescent color under both long and short wave ultraviolet. Phosphorescence should also be looked for. While some minerals are brilliant and certain in their fluorescence, it may be difficult to be sure that other mineral specimens fluoresce, even under the best conditions – at home, in total darkness. This is particularly true of weakly fluorescent minerals, or those which naturally fluoresce blue or violet. Both short and long wave lights pass some visible light particularly in the blue and violet portions of the spectrum. The visible blue-violet light which is passed by the filter will reflect from many mineral specimens, producing a blue, violet, or purple appearance which is difficult to distinguish from a true blue or violet fluorescence. If there is some question about the response, there are a few tests that can be applied if short wave ultraviolet is involved. Since short wave ultraviolet will not pass through glass or a thick sheet of clear plastic, a sheet of glass or plastic can be held between the ultraviolet light and the mineral in question. If the blue or violet response disappears, it must have been due to short wave ultraviolet, which is blocked by the glass or plastic, rather than to bluish light which passes through. For this purpose, it is important to use a glass or plastic which itself does not fluoresce. This is harder to find than one might think, as many transparent plastics and most glasses fluoresce under short wave ultraviolet. Another somewhat less certain test, but one applicable to both short and long wave is the following. Select a nonfluorescent mineral which appears to have the same color and texture under visible white light as does the mineral specimen in question. These two minerals will then have about the same degree of reflection of visible violet. Compare these two under ultraviolet. If the mineral specimen being examined produces noticeably more blue or purple light than the nonfluorescent comparison piece, the additional output may be due to fluorescence.

Specimens should be trimmed to a size and show of color which allows the fluorescence to be seen to best advantage and in the most attractive way. For more easily found fluorescent material, trimming can be based on aesthetics – the best balance and distribution of color and intensity. For rarer minerals, the desire to not discard a scarce material must dominate over aesthetics. With such material, care should be taken not to discard rare mineral material for the sake of improved appearance.

Saws are often used for trimming, but where fluorescent minerals are concerned, they must be used with caution. Most saws use either an oil or a detergent lubricant. These liquids are very difficult to remove from a specimen completely, and as they may also fluoresce, they can spoil the natural fluorescence of a specimen. For this reason also, detergents should not be used to clean fluorescent minerals. Ordinary hand soap applied with a clean brush is usually sufficient to remove loose dirt.

Some crystal specimens have been freed of a surrounding matrix, usually calcite, by means of acid which selectively dissolves the matrix. This practice should be approached with caution as far as fluorescent minerals are concerned. Acids often react subtly with elements present to produce a thin coating of altered chemistry which may fluoresce of its own, having little to do with the fluorescent properties of the mineral being treated.

As the collection grows, the fluorescent collector will need to organize and display his growing mass of accumulated material. He will probably segregrate his most interesting and showy specimens in a display collection, and the remainder in a reference collection. For smaller study and reference specimens, a chest of low drawers or a steel "30 drawer" cabinet is ideal for storage. Here, material can be kept in small open boxes or box lids. Cardboard boxes usually fluoresce blue-white under ultraviolet light due to "brighteners" – fluorescent dyes – added to the paper of which the boxes are composed, and the same is true of the cotton which is usually found in a mineral box. This bright fluorescence distracts from the subdued fluorescent response of minerals. For this reason, the cotton should be discarded, and boxes holding fluorescent minerals should be sprayed with a flat black paint. Similarly, labels which are glued on the specimen or kept with it should be of a nonfluorescing paper or card stock. Nonfluorescing paper will provide enough visible reflection under ultraviolet so that handwritten or typed information can be clearly read in the dark. Labels should be written in one color, preferably black. Fluorescent ink should be avoided, while a variety of fluorescent inks distract from the fluorescence of the minerals themselves. The label should contain the name, or names, of the minerals of which the specimen is constituted. Most important, the place of origin should be stated. This should be given with as much precision as knowledge allows. The best practice includes the particular location within the mine where the mineral was found, if known. Location is, if anything, more important than mineralogical identification. Uncertain identification can always be corrected later, if need be, through the aid of chemical, optical, or X-ray diffraction analysis. However, location, once lost, is usually lost forever. Lack of location on a mineral specimen tends to sharply decrease the value or interest of a specimen.

The material in a reference collection can be examined and studied conveniently with the use of a hand held ultraviolet light. A short wave and long wave, or dual band, light will be needed. The display collection can also be viewed with a hand held light, but most collectors will want to set up a showier display which can be illuminated as a whole. For long wave fluorescent minerals, "black light" type fluorescent tubes are a convenient and comparatively inexpensive source of ultraviolet. These lamps, made for commercial displays or home fluorescent poster display, are identical in size to the ordinary visible light fluorescent tubes used in offices and elsewhere, and the same light fixtures are used.

Short wave minerals present a greater display problem since short wave lights are expensive and there is no lower cost substitute. Also, these lights are subject to solarization after a few hours of use in a display application unless the newer type of short wave filter is used. For these reasons, a moderately priced hand held light, moved over the specimens one after another, may be the best compromise for many.

A word of caution: if the minerals are examined frequently under short wave light, it is best to wear glasses for protection. Eyeglasses will do very well. If one does not wear glasses, clear plastic work goggles will do, and these can be obtained at a hardware store for a few dollars. It is also wise never to look directly into either short or long wave light.

5
Key Collecting Localities and Their Fluorescent Minerals

Gold is where you find it. This phrase expressed the gold prospector's occasional perplexity that there seemed to be neither rhyme nor reason in where gold was to be found. However, the experienced prospector knew, in fact, that there was a pattern in where gold might be found, and he knew also that he might expect certain other minerals in close association with the precious metal.

An experienced mineral collector knows likewise, that certain minerals and certain mineral groupings are to be expected in certain geological settings. Thus, he will anticipate zeolites in basalt and diabase rocks, and gemmy silicates such as beryl in pegmatites, but he would be somewhat surprised to find zeolites in pegmatites, and astonished to find beryl in a basalt.

The occurrence of fluorescent minerals follows the same patterns which are seen in other minerals. The fluorescent collector will need to understand such patterns if he is to know what fluorescent minerals may be expected in different geological and mineralogical settings. He may compare the locations described here with similar geological settings elsewhere, perhaps closer to home, and may possibly find similar fluorescent minerals to be present.

In addition to providing a guide to collecting, this knowledge is also useful as a guide to mineral identification. In view of the bewildering variety of minerals and their many habits and disguises, the setting from which they come is often the first and best clue to identification. It allows one to narrow the possibilities. Taken together with the other clues – color, luster, hardness, specific gravity, and crystal form if the mineral is outwardly crystalline – the geological setting often provides a reasonable basis for identification. Fluorescent response provides an additional dimension of identification, and the fluorescent color tables in Chapter 10 may be consulted.

Minerals are disseminated thinly throughout the earth's crust, but commercial exploitation requires those unique concentrations variously called ores or mineral deposits. Here, if location, cost of extraction, and other factors are right, quarries are opened or mines are sunk in search of one or more constituents of the ore or deposit. These locations are also the primary source of minerals for the collector, including the fluorescent mineral collector. It is true that one can find an occasional "wild" specimen, one not brought forth from commercial workings. Thus, one may find an occasional specimen on the surface of the desert or in some unexploited rock outcropping, but most successful collecting will be accomplished on the dumps of an active or dormant mine or quarry. Or, more frequently, the specimen will be purchased, but the ultimate source remains the same, a commercial working exploiting some particular geological and mineralogical environment.

In the following discussion, some of the more important environments or settings in which fluorescent minerals may be found are described, and the possible or likely fluorescents which may be found in such settings are reviewed. A brief sketch of the current view of the origin of the mineralized bodies is given. In most cases, there is dispute among geologists about such matters, and the origin of various mineral bodies is and has been a source of controversy in the geological community going back to the beginnings of geology as a modern science. However, in its broad outlines, what is said here is essentially in line with current beliefs, but refinement in theory and change in the viewpoint of geological scientists can be expected as time goes on.

PEGMATITES

The large chunky blocks of feldspar, pods of mica, and occasional gemmy tourmaline, beryl, or topaz, are typical minerals of pegmatites and are probably familiar to most collectors. The searcher after rare minerals turns to the pegmatites for certain phosphates, rare earth minerals, and a number of other minerals which tend to be peculiar to one or a few pegmatite occurrences. The fluorescent collector too may find interesting material in pegmatite outcroppings and dumps. As with nonfluorescent minerals of the pegmatites, some fluorescent minerals are common at many pegmatite locations while others are rare one-location minerals.

Pegmatites are sometimes seen as thin veins of more or less uniform composition running through a gneiss or other host rock. Such pegmatites are rarely the interesting ones from the collector's viewpoint. Rather, it is the large massive outcroppings, tens to hundreds of feet in extent, that usually contain the interesting mineralization. These are quarried for industrial mica, for feldspar used in scouring powders or high temperature ceramics, or for rarer materials such as beryllium or lithium minerals.

Such large feldspar bodies are found in the roots of ancient mountains of the eastern United States, in Vermont, New Hampshire, Connecticut, Pennsylvania, through Virginia and the Carolinas to Georgia. In the West, well-known mineral rich pegmatites are found in South Dakota, Colorado, New Mexico, and California. The occurrence of pegmatite is certainly not limited to these locations, however.

The origin of pegmatites is one of those subjects which has been in long dispute in geological circles. The theory of origin is interlinked with the theory of origin of granite – one of the most controversial subjects in geology – and the theory of origin of great metal deposits. To enter a discussion of pegmatite origins, it is useful to first discuss the origin of granite since pegmatites are believed to originate as a consequence of the formation of granite.

The Sierra Nevada of California and the White Mountains of Vermont contain examples of large outcroppings of granite-like rocks (including granite, granodiorite, and quartz monzonite) of the kind found in various mountainous regions of the country. Large bodies of this type, many miles in extent, are known as batholiths. Smaller such bodies are in the form of stocks or dikes. These masses are formed well below the surface from high temperature liquids. They now are found on the surface since layers of softer sedimentary or metamorphic rocks once above them have been eroded away. Because they are thought to have moved to their present location from below and forced up through surrounding rocks, these batholiths and stocks are referred to as "intrusives."

The depth of formation, the source of the forming material, and the nature of the liquid from which these bodies solidified have been at the center of one of the major controversies in geology for over a century. It has not been fully settled to this day. The conventional view is that the liquid is a high temperature melt – a magma, but one fundamentally different from basalt in composition and origin. If the melt was not close to granite-like in composition to start with, it becomes so as the melt cools, and minerals not typical of a granite crystallize and separate from the melt. Or, the origin may not be a true melt, but a chemical alteration and recrystallization of sediments under high pressure and temperature near the melting point.

The high temperature and pressure required may originate in the crumpling and folding of the earth's crust during mountain building. However, the modern theory of plate tectonics has provided another interpretation. Using this view, it has been suggested that granites originate from ocean sediments carried down to great depths by an oceanic "plate" descending below the edge of a continent. At sufficient depth, the water rich sediments are melted, due to the pressure at depth. The melt then works its way to the surface, perhaps aided by buoyancy, fluidity, and preferred pathways due to deep faults. This viewpoint is also exploited in a present day theory of the formation of certain metal deposits, which will be discussed later.

We now assume that the melt has risen to a region close to the surface. It has begun to solidify due to reduction of pressure and loss of heat. The not yet hardened granite contains a small amount of water, perhaps a few percent. As the granite continues to cool, crystals form and water is ejected, since it has no place in the composition of quartz, feldspar, and other crystalline constituents of the granite. This water concentrates in the yet unsolidified portion of the great high temperature mass of forming rock. Other elements, including fluorine and chlorine, phosphorus, boron, lithium, uranium, and the rare earths, similarly concentrate in this water rich melt which also still contains in solution the elements needed to make quartz and feldspar. Because of the increased concentration of water and a few of the other constituents, this remaining material is now much more fluid and mobile than the parent melt. It is thus able to move about very easily under the influence of pressure of surrounding rock. As the outer portions of the great mass of granite solidifies and occasionally cracks and fractures, this mobile liquid is rapidly injected into the fractures. It may also leave the granite mass altogether and enter the surrounding host rock. Here it slowly cools and hardens as a pegmatite.

Due to the low viscosity of the forming liquid, the pegmatite contains crystals of much larger size than that of the parent granite, and due to the concentration of elements rejected by the parent granite, the many interesting minerals for which pegmatites are famous are formed. These are not found uniformly through the pegmatite. Pegmatites are often zoned. Along the walls of the pegmatite, an interaction with the host rock produces a fine grained mix of feldspar, quartz, and possibly some tourmaline, apatite, beryl, and garnet. Toward the inside, these constituents become larger in size, and this is the main zone for tourmaline, mica, beryl, and rare earth minerals also. Further to the center, the constituents are various feldspars, quartz, biotite, muscovite, large beryl crystals, and columbite-tantalite. But if lithium is present, amblygonite, lepidolite, and spodumene are also included. Finally, the same minerals may appear in the core of the pegmatite, but quartz will be the dominant mineral here.

The Strickland quarry near Portland, Connecticut, for example, is a typical or representative pegmatite, and most of the aforementioned minerals are found there. Several interesting fluorescent minerals are among them, including:

apatite
autunite
hyalite
microcline

It is often possible to find fluorescent apatite, hyalite, and microcline in combination. Apatite is manganese activated. The autunite and the hyalite are activated by uranium (uranyl), probably originating from uraninite. The microcline is probably activated by europium.

Other pegmatites are much richer in lithium minerals and produce a distinctive suite of fluorescent minerals. Depending on which lithium pegmatite it is, it may provide:

eucryptite
kunzite
spodumene

In other pegmatites, other fluorescent minerals are found, including:

beryl
brannockite
fluorite
herderite
tourmaline
uralolite
zektzerite
zircon

Fluorescent zircon is not uncommon in pegmatites, while the other minerals mentioned are generally rare and may be unique in some cases to one pegmatite. In Chapter 6, Part B, fluorescent minerals from outside the United States are discussed. A number of pegmatite minerals are included, an indication of the richness and variety of the fluorescent minerals furnished either commonly or rarely by pegmatites.

METAL ORE DEPOSITS

The mines which yield copper, zinc, lead, and other metals for industrial use are prolific producers of mineral specimens in bewildering variety. Many fluorescent minerals are found among these specimens. While a number of such mines can be found in the eastern part of the country, the great majority are found in the West, in what is one of the greatest metalliferous zones of the world.

There would seem to be little in common in terms of the origin of such metal mineral deposits in the West, distributed apparently randomly over the map and made up of different combinations of major as well as minor minerals. However, an underlying pattern of origin is believed to exist.

The earlier discussion on pegmatite formation dealt with granite-like intrusives. Maps of the western part of the United States have been made showing the location of such intrusive outcroppings. To the west and north, the maps show the great granitic mountain ranges – batholiths in our terms. In the Southwest, smaller granitic bodies predominate. Now if the locations of the major metal mines which yield copper, gold, zinc, and lead (often the same mines produce all of these metals in varying degrees) were to be superimposed on the map of the intrusives, a striking coincidence of location of metal concentration and of intrusives would be evident. There are exceptions. Some metal deposits would appear with no obvious granite body nearby, but in such cases, a nearby granite body may be hidden well below the surface.

The close proximity of metal ore deposits to intrusive rocks leads geologists to the inevitable conclusion that the formation of these metal ores is closely related to the existence of the nearby intrusive. While that point is generally agreed upon, there is less agreement on what the relationship is.

Some few believe that the relationship between intrusive and metal ore is only that they share a fault system by which they have independently found their way toward the surface as liquids or melts. Indeed, a map of major faults shows that both the intrusive and the metal ore deposit are often located at such faults. In Nevada and northward where faulting is still seismically active, the metal deposits are often found close to the location of earthquake occurrences. However, most geologists believe that the metal deposits originate from the intrusive granite mass. Beyond this point, there is still considerable disagreement on the details. In the most common view, the metals concentrate in the hot, watery solution which remains after most of the granite mass cools and solidifies, much as in pegmatite formation, though now the percentage of water is much greater. This metal bearing water, probably a salt rich brine, is forced into faults and fissures in the upper part of the cooled mass of rock, or out into faults in the surrounding rock where the metals are deposited as sulfides. In an alternative view, the minerals may originate from high temperature water from the outside which moves through the almost solid granite melt, leaching it of metals which are then deposited elsewhere. Either way, the fissure deposits become veins of ore and the home of many interesting minerals.

As far as the great copper deposits of the West are concerned, an alternative theory has been proposed exploiting the theory of plate tectonics. According to this view, at the time of formation of these ores some 30 to 100 million years ago, the western portion of North America was underthrust by the eastern Pacific Ocean crust, or plate. At the line of underthrust, the ocean crust or plate began a deep and sloping descent beneath the continent. This ocean crust was rich in copper. At a point in the descent where the depth was sufficient to provide melting heat, the ocean crust partly melted. The copper rich melt, or possibly a salt brine, rose within the continental crust to join or to form the granitic magmas from whence the copper ores later separated. Possibly the other metals – zinc, lead, gold, tungsten, and others – owe their origin to the same source. This theory is thus a variant or an extension of a similar theory of pegmatite origins.

The metal deposits of the West thus may have a common origin in granite-like magmas or melts whose present disposition is in the form of outcropping intrusive rocks. The metals were separated from the magmas by rising hot waters carrying away metals in solution. To move back in time one further step, it is possible that some, or maybe all, of the metals may have originated from ocean crust. Beyond these shared features, the ore deposits are individualistic to some extent, due to differences from place to place in the mixture of metals carried in solution, different liquid temperatures and pressures, different conditions of deposition, or different history after deposition.

What present evidence exists to support the view that the metal ore deposits were transported and deposited by hot, watery solutions? In the lower levels of the mines in the Comstock Lode, miners were often impeded in their work by the outrush of hot, scalding, mineral bearing waters from below. The mineral deposits of the Comstock are probably of recent origin geologically speaking, and the nearby intrusive which is the probable origin of the ore is thought to be still hot at depth. Thus, human activity may have intruded on this metal deposit while it was still forming.

Perhaps the simplest of the mineralizations of interest is that of scheelite. A particularly simple example is presented by the deposit at Atolia, California. Here, scheelite has been deposited as a vein filling in faults. Close by to the north, these faults contain gold, and some mines have produced both gold and scheelite. Both have been deposited from hot, mineral rich solutions, that is, by hydrothermal deposition, originating in a magma which may have also fed the extinct volcano nearby. Silica in solution was probably the carrier for both gold and tungsten since these metals both have an affinity for silica when in solution. The silica has joined these metals in the form of quartz. Calcium for the formation of scheelite was probably supplied by calcite or limestone formations in the rocks through which the hot fluids moved.

The fluorescents to be found at Atolia are:

scheelite
hyalite
calcite

The hyalite is probably a late addition, carrying traces of uranyl as an activator. The calcite is a gangue, showing the characteristic orange fluorescence of exposed desert calcites.

Near Oracle in Arizona, scheelite is deposited within a large mass of calcite. This calcite is an intensive red fluorescent, indicating the presence of lead and manganese.

Far better known for economic importance and for choice minerals are the many mine localities which have produced copper, zinc, lead, and in many cases, originally gold or silver. At these numerous sites, particularly in the West, heated water carrying metals in solution (hydrothermal fluids) has deposited sulfides and, to a lesser degree, arsenides or antimonides. Apparently fluorine was often present in the hot solutions, reacting with limestone in the region of deposition to form fluorite. Such fluorite often fluoresces weak to intense blue, indicating the presence of europium. Such blue fluorescing fluorite is a likely indicator of nearby heavy metals, and the europium presumably was contained in the original ore bearing liquids. Similarly, red fluorescing calcite may be present, and this too is an indicator of nearby heavy metals. In some deposits, the two are found together providing nice two-color fluorescent specimens.

The sulfides, arsenides, and antimonides are a dull fraternity of minerals as far as the fluorescent collector is concerned. With the exception of sphalerite, they do not fluoresce. They are also of no exceptional interest to the general mineral collector. However, when erosion has opened the top of the ore body to the atmosphere, a wonderful mineralogical transformation takes place. Surface water or rain will carry dissolved oxygen down into the ore and work chemical changes. Iron, lead, and zinc sulfates will be produced. If certain bacteria are present, sulfuric acid will be produced rapidly by further oxidation of the iron sulfate. Dripping down the vein, this acid will convert other lead, zinc, or copper sulfides to sulfates. The entering water may also contain carbon dioxide in solution as carbonic acid, and this will react with the sulfides to produce carbonates. Some of the sulfuric acid produced through the oxidation of sulfides will pass in contact with limestone and additional carbonic acid will be produced, thus furthering the production of carbonates.

One result of this oxidation is that limonite will now stain the host rock in the upper portion of the vein a rusty brown. This rusty region is called a gossan and will contain numerous small cavities where the now removed sulfides once resided. The gossan may contain gold; it may also contain anglesite and cerussite which, being insoluble in water, are usually found close to the location of the now vanished galena from which they formed. Further down, other sulfates and carbonates, oxides and hydroxides, and also arsenates and silicates

Figure 5-1. The Mammoth mine at Tiger, Arizona, around 1914. The mine is famous for the large variety of common and rare minerals found, including a number of fluorescent minerals. The multitude of other metal mines throughout the West produced supergene minerals, including fluorescents, though few such mines matched the variety found at Tiger. (Photo courtesy of the Arizona Historical Society, Tuscon, Arizona)

are produced and deposited. It is these inhabitants of the so-called oxidation zone that provide the collector with the many colorful, attractive, and often rare crystals for the collection. Many of these may fluoresce, as will some of the accompanying gangue minerals.

The number of such mines in the West is seemingly limitless. These include mines well known to crystal specimen collectors: the Queen mine at Bisbee, the Mammoth mine at Tiger, the King of Arizona (Kofa), all in Arizona; the American Tunnel and Bird Camp in Colorado; the mines of the Kellogg district in Idaho; the mines at Darwin, California. These are but a few. In the East, such metal deposits are significantly rarer – the mines at Phoenixville, Pennsylvania, being typical.

The fluorescent minerals which may be found at such mines include:

adamite	fluorite	smithsonite
anglesite	hemimorphite	willemite
aragonite	hydrozincite	wulfenite
barite	leadhillite	(zeolite minerals)
calcite	matlockite	(uranium minerals)
cerussite	pyromorphite	

Several fluorescent mineral combinations are encountered including calcite and fluorite; calcite and willemite; fluorite and barite; and hydrozincite and calcite. Certain others of these fluorescent minerals are scarce to rare in this country, including adamite, leadhillite, matlockite and pyromorphite.

Hydrothermal ore deposits in which uranium is the primary metal of economic interest are unusual in this country. The Marysvale district in Utah is such a deposit and has produced several interesting fluorescent minerals. The veins which contain the ore are in close proximity to the intrusive granitic rock which was their source some ten million years ago. The ore is pitchblende, which is not a fluorescent. As in the other metal mines, it is the effect of oxidation and alteration which produces the interesting fluorescent minerals. The fluorescents of Marysvale include the following uranium minerals:

autunite (meta)
schroeckingerite
uranophane
uranopilite
zippeite

BASALTS

The zeolites and other accompanying minerals furnish many a collection with a variety of attractive specimens. While they are found in other environments, the most spectacular specimens are found in certain basalt and diabase formations. Of these formations, the Watchung Mountains and Palisade sills of New Jersey have probably provided the largest number of fine specimens in this country.

The Watchungs run southwest from Paterson, New Jersey, to Somerville, New Jersey, in three separate but parallel chains known as the first, second, and third Watchungs. The first is the easternmost. These mountains are the exposed edges of thick blankets of lavas, poured upon the surface of the earth in three episodes almost 200 million years ago.

The Palisades, about 15 miles east of the first Watchungs, are diabase of apparently the same source composition as the Watchungs, formed at about the same time. Like the walls of a great fortress, hence their name, the Palisades form tall columns on the western bank of the Hudson, opposite New York City.

Vast sheets and flows of lava such as the Watchungs and Palisades are not volcanic discharges. In concert with the theory of plate tectonics, the Watchungs and Palisades are thought to have formed as extrusions through long fissures rent in the earth's crust at the time that the vast supercontinent of Pangaea broke up. Parts of the first

Watchungs may have been formed by lava which flowed into lakes, since the formations are billowy in form – pillow lava – and permeated with open cavities and bubbles. It is in these openings and bubbles, and in other available fissures in the basalts and diabases, that the zeolites and other minerals of interest have been formed.

The mode of formation of minerals within these cavities has been a matter of some controversy. Sodium, calcium, aluminum, silicon, sulfur, boron, and other elements required in the formation of the resident minerals are present in the lavas. In one view, these elements were carried as soluble minerals by water flowing upward under pressure through fine fissures in the hardened lava. These were deposited within the cavities and bubbles in the lava. Early formed minerals were sometimes replaced by later forming ones, and entire suites of minerals were assembled.

The zeolites make up the greatest number of minerals to be found, but calcite, prehnite, and numerous others are also found. The order of mineral formation started with the sulfates such as glauberite and anhydrite, followed by prehnite, datolite, and pectolite. The zeolites and calcite then formed.

While some of these minerals, particularly the zeolites, are reluctant and weak fluorescents, the other accompanying minerals are often attractive fluorescent specimens and encourage the search for good material. The minerals found in the Watchungs or Palisades which may fluoresce include:

Zeolites	**Non-zeolites**
analcite	apophyllite
chabazite	calcite
heulandite	datolite
laumonite	pectolite
mesolite	prehnite
natrolite	thomsonite

The red fluorescence of some calcites from these formations indicates the presence of lead and manganese, while the more subdued whitish fluorescence of other calcites and the zeolites perhaps suggest the presence of rare earths. Such materials are not lacking in these basalts, though they are disseminated rather than concentrated as in an ore deposit.

Basalts in other regions may provide other fluorescents, and the minerals listed above can be found in environments other than the basalts.

THE FLUORSPAR BELT

The fluorspar (fluorite) mines of southern Illinois have produced so many fine crystalline fluorites over the years that such specimens are known and probably recognized by sight by most collectors. The generous size of the crystals, the faces sometimes marked by raised square embellishments, and the purple or purple-red color, or occasionally a pale blue color hiding a phantom of yellow, are characteristic of specimens from these deposits. The fluorspar deposit from which these specimens are taken extends from Pope and Hardin counties in southern Illinois across, or rather below, the Ohio River into Livingston and Crittenden counties in Kentucky. It is one of the largest and most important fluorspar deposits in the world, worked for over 100 years for fluorite for industrial purposes. To the mineral collector, however, two localities stand out. These are Rosiclare and Cave-in-Rock, 17 miles apart on the Ohio River in Hardin County. It is from mines at these locations, and Elizabethtown in between, that the best mineral specimens have come.

For the fluorescent collector, it is not the fluorite from these locations which is of greatest interest. Better fluorescent fluorite is found elsewhere. Rather, the fluorescent collector will be particularly interested in other minerals which will be discussed shortly.

The geological origin of the fluorite and other minerals of these mines is understood in an approximate way. The theory of origin of these deposits, like any geological theory, must utilize the observed facts of the situation and must also explain these facts. First, the region is one of the most intensely faulted regions of the United States. This fault system has a predominantly northeast to southwest orientation, and the faults appear to merge southward along the Mississippi with faults of the New Madrid system. The famous earthquake at New Madrid in 1811 was probably associated with movements on these faults.

The fluorspar district also shows evidence of magma intrusions at one or more times in the distant past. One present evidence of this is the presence of metal sulfides here and there in the fluorite deposits.

In consequence of these observations and others, the following rough sketch of the origin of the minerals has been proposed. Perhaps 100 million years ago, the ancient limestone beds which existed in this area were ruptured and faulted by the upward pressure of a major underground intrusion of molten rock and its subsequent retreat, which first broke the limestone strata and then allowed sections to slump. Then this intrusion of molten rock, or perhaps a later intrusion, produced a fluorine rich hydrothermal fluid which forced its way up into these faults. It appears that this fluid contained salt and was thus a brine of sorts. It is not surprising, therefore, that the fluorite belt contains occasional deposits of metallic sulfides, since hot brines, particularly if fluorine bearing, are powerful transporters of dissolved metals.

On reaching the limestone strata, the hot fluorine bearing fluid reacted vigorously with the limestone, dissolving it along the weak and permeable cleavages between strata, and combined with the calcium in the limestone to form fluorite, replacing the limestone with fluorite and carrying additional dissolved fluorite onward to be deposited in veins elsewhere within the faults. These veins, typically 1 to 15 feet in width, are the major source of industrial fluorspar for which the district is famous.

At some locations, particularly at Rosiclare and Cave-in-Rock, this process left open cavities behind. Some of these cavities were 3 or 4 feet wide. Dissolved fluorite entered these, perhaps drop by drop from the roof of the cavity. These solutions fed the slow growth of the fluorite crystals which have since been found in profusion for the sake of generations of collectors.

Calcite, barite, and other crystallized minerals occasionally found with fluorite crystals are of later deposition. These may have had an origin in descending warm waters which carried these minerals in solution and deposited them within the cavities often over the surface of fluorite.

Some of the minerals from these fluorite mines are of particular interest to the fluorescent collector, including:

alstonite
barite
barytocalcite
benstonite
calcite
fluorite
strontianite
witherite

The fluorite is an occasional fluorescent. The weak violet fluorescence of some of these fluorites suggests the remote origin of this material in a magma, the likely source of an attenuated rare earth activator. However, somewhat more commonly, a yellow fluorescence is seen. This has been attributed to an organic activator, and indeed, tar-like inclusions are found within such fluorites. Organic materials are certainly present in the fluorspar ores, and oily or asphaltic oozes are encountered in the Illinois mines. In one case, a cavity some 30 cubic feet in volume was encountered, half filled with viscous, tarry material.

The alstonite, barytocalcite, and benstonite are scarce or rare, but the other minerals appear in the hands of dealers with reasonable regularity. The calcite, barite, and strontianite from these locations are attractive fluorescents and superb crystals. The witherite is unmatched as a fluorescent by witherite from any other source.

CENTRAL UNITED STATES

In the middle of the United States there is a region in which sedimentary rocks dominate. These include predominantly limestones, but also shales and sandstones. This region extends westward to the prairies and eastward to the edge of the Appalachian Mountains. Where the Appalachians swing northeast into New England, the region follows into northern New York State. To the south, it extends into eastern Tennessee and westward into Arkansas and into west Texas.

Geologically, this region has been comparatively stable and uneventful for hundreds of millions of years. It has been little affected by the repeated periods of major earth movement and mountain building which have affected the regions to the east and the west. Being low lying and relatively flat, it has experienced several episodes of incursion from the sea in ancient times, and shallow seas have covered the region for substantial periods.

The southern portion of this region contains numerous lead-zinc mineral districts. These metal ore deposits may have a common origin, but a satisfactory theory of their origin has not been worked out. However, it is the northern portion of the region which offers the greatest interest from the viewpoint of mineral fluorescence.

This northern zone from Iowa through Illinois, Indiana, Ohio, western and central Pennsylvania, and northern New York contains a distinctive group of minerals, with apparently much in common from place to place within the zone. These minerals in diverse locations may ultimately be found to have a common geological origin of their own. While such a shared geological origin is not established, they are treated together here because of the similarity of the minerals found, because of the fact that they occur in limestones throughout this wide belt, and because of the similarity of fluorescent response of the minerals involved. For example, there are found in western Ohio the well-known honey colored fluorites of Clay Center and Weston. These often occur with crystals of barite or celestite. The fluorite often contains cubic phantoms within, and these are the center of a bright yellow-white fluorescence due to an organic activator. To the west in Iowa, virtually identical fluorite is found in Black Hawk County. Again, the fluorite contains internal phantom crystals which are the center of the major portion of the fluorescence. To the east in Indiana, similar fluorites are found in Fort Wayne with a whitish fluorescence, but occasionally in some specimens showing a phantom with yellow fluorescence. To the south, a similar fluorite is found in Danville, Kentucky, accompanied by fluorescent barite. This fluorite shows an internal phantom with good whitish fluorescence, but here the barite fluoresces weak red rather than white, which suggests a trace of manganese in its composition. In Ohio, the presence of celestite was mentioned. Celestite is also found in the limestones northward in nearby Maybee, Michigan, and eastward into New York at Chittenango Falls.

A general resemblance seems to exist between these deposits and those of the Illinois-Kentucky fluorspar discussed previously, which is on the southern edge of the region described. Thus, the Illinois fluorites contain internal phantoms which fluoresce yellow due to an organic activator. The Illinois mines also contain barium as both sulfate and carbonate (barite and witherite), but contain strontium primarily as the carbonate in the form of strontianite, rather than the sulfate, celestite.

Possibly, the Illinois-Kentucky fluorspar zone scheme of mineral origin pertains also to these other deposits spread as they are from Iowa to New York. In brief, this would require deep intrusive magmas releasing fluorine upward into the limestone to form fluorite, and downward leaching of disseminated sedimentary deposits of barium and strontium minerals redeposited as barite, celestite, or strontianite. Support for this view, however, is not particularly strong, at least as far as the origin of fluorine from a deep magma is concerned. Little evidence exists outside of the Illinois district for the deep intrusives, and these deposits do not contain much in the way of sulfides of zinc and lead which would be a supportive sign of ascending hydrothermal fluids.

Thus, it cannot be taken for granted that this vast sweep of territory described here represents a common district as far as the origin of these minerals is concerned. A more thorough comparison of minerals from these locations is required, and a unifying theory of origin would have to be established. Then, the peculiarities of each site would need to be accommodated. As an example, elemental sulfur is found with blue celestite and calcite at Maybee. This sulfur may be formed by bacteria acting on earlier sulfates which they may completely reduce to sulfur. Interestingly, this suite of minerals from Maybee resembles those of the much more famous mineral deposits of Agrigento in western Sicily, and the origin may be substantially the same.

Whether or not these locations may be properly grouped together on a scientific basis, they may be grouped together on the more practical basis that similar minerals occur. Thus, in the limestones of this extensive region, the following fluorescent minerals may be found:

barite
calcite
celestite
fluorite
strontianite

Of these, the fluorite is usually the best and most consistent fluorescent, but large crystals of celestite and small tufts of strontianite also provide attractive fluorescents.

THE TRANSFORMED LIMESTONES

Substantial portions of the land are covered with limestone, originally formed perhaps by lime secreting algae or from the limy skeletons of simple animals in the sea. In those places where the limestone has later been subject to high temperature, as in the vicinity of a hot magma or a hydrothermal intrusion, a substantial alteration in the limestone will have taken place. It is consolidated and crystallized into beds of crystalline calcite or marble. Organic matter in the limestone, if present, is driven out by the tight fit of atoms in the calcite crystal structure. Such organic matter may be carbonized by heat and may appear as a graphite disseminated through the calcite bed. Also, if silica in the form of sand or volcanic ash was originally present, and if the temperature was high enough, many interesting calcium silicate minerals may have formed. If the limestone also contained magnesium, still more interesting and rare minerals may have formed. Finally, if yet other elements were present, the mineral forming possibilities multiply rapidly. Among other minerals of the metamorphosed limestones, there may often be found a generous number of fluorescent minerals.

Many collectors are probably familiar with, or have heard of, the minerals of the Crestmore quarries three miles northwest of Riverside, California. Here in an unusual operation, the descending limestone layer is followed underground by mining in pursuit of high grade material for use in the manufacture of cement. Such pure limestone is of no great interest to mineral collectors, but elsewhere, on the surface in what is called the Commercial quarry, it is another story. Here, an ancient event has profoundly altered the limestone and created many of the minerals for which the location is famous.

The total number of mineral species thus far recorded at Crestmore rivals the count of species found at Franklin, New Jersey. The majority of these minerals consist of calcium and magnesium combined with silicate, carbonate, or hydroxyl groups in various ways. The origin of this vast assemblage lies in the ancient intrusion of a hot, granitic magma into magnesium rich limestone beds. The resulting heat acted on and transformed the limestone, and the highly fluid magma may have injected other materials into the limestone as well. Altered rock formed around the hot, chemically potent intrusive, and is now found in a series of rough and irregular zones of alteration. Nearest the intrusive mass, a rim of rock contains grossular garnet, wollastonite, and dioposide. This is surrounded by a second zone containing idocrase, garnet, monticellite, wilkeite, tobermorite, spurrite, and merwinite. Outside of these zones, there is a zone of monticellite within which is found spurrite, tilleyite, scawtite, and spinel. Finally, outside of these predominantly silicate minerals, crystalline layers of calcite are found in a bed about 500 feet thick. The color of this calcite is very unusual, being blue to blue-gray, and is known as the Sky Blue limestone.

The presence of monticellite, spurrite, and merwinite suggests that the magma intruded the limestone while the limestone was near the surface, since these minerals form at high heat but at comparatively low pressure. Also, the zone of transformation may have had access to the surface through fissures, since the venting and removal of carbon dioxide derived from heating of the limestone encourages the formation of these silicates.

Franklin collectors will be interested to know that among many other minerals, chondrodite, ettringite, fluoborite, xonotlite, and nasonite have been found at Crestmore. At Franklin, the first three are fluorescent, and at one time, nasonite was thought to be fluorescent. It is unfortunate, therefore, that the Crestmore minerals have not been the subject of the thorough study from the fluorescent viewpoint to which Franklin minerals have been subjected. Such study would provide a useful comparison of minerals which are common to the two locations. It would also probably reveal that many more Crestmore minerals are fluorescent than are listed here.

The fluorescent minerals of Crestmore include:

apophyllite
aragonite
calcite
deweylite
forsterite
foshagite
gyrolite (centrallasite)
harkerite
hyalite
microcline
plombierite
scapolite
strontianite
thaumasite
tobermorite (crestmoreite)
wollastonite
xonotlite (jurupaite)
zircon

Fluorescent minerals said to be prehnite, scheelite, szaibelyite, and uranophane have also been found. An investigation of collected material is likely to reveal many more fluorescents among the minerals of Crestmore.

In the vicinity of Franklin, New Jersey, the Franklin marble, a transformed limestone, is a prolific producer of minerals, a number of them fluorescent. Many active or dormant quarries have worked in this marble, producing limestone primarily for agricultural purposes. Occasionally, a quarry opens to collectors, and at such times, several hundred collectors from half a dozen states may be present. Few return home empty handed. The Franklin and Sterling Hill ore bodies are located within the Franklin marble, and minerals of the marble are found with the ore. Such minerals are loosely considered to be "Franklin" minerals when they originate from the Franklin or Sterling mines, but they are found in many other places throughout the length of the extensive Franklin marble bed and thus should be accorded consideration separate from those of Franklin.

The origin of the Franklin marble is to a great extent lost in geological history since it began well over one billion years ago. At that time, limestones were deposited in warm seas as limy oozes. These limestones were interbedded with silica, possibly a sandy sediment, ash from volcanic activity, or clays. At a later time, these strata were raised above sea level. Subsequently, perhaps 800 to 900 million years ago, the limestone was subject to high compression, folding, and heating over an extensive region in conjunction with major geological changes, perhaps mountain building. This changed the limestone to crystalline marble or calcite, the silica and clays to gneiss, and produced the many silicate minerals now found within the marble, such as diopside and various amphiboles. At some time in the history of this formation, probably before heating and compression, fluorine and boron entered the formation since many of the minerals including norbergite, edenite, and fluoborite contain fluorine or boron. Such elements might have originated from saline lakes during the rise of the land from the sea, or they may have derived from magmas at some depth. At some time after metamorphosis, pegmatites and other melts intruded, producing feldspars and scapolite at the interface with the calcite.

Through this complex geological history, or through a similar sequence perhaps differing in some details, the present Franklin marble was formed. It is found in a narrow belt starting in the south in the vicinity of Sparta, New Jersey, and ending some miles north, near Edenville in southern New York. Similar marble is found northward in St. Lawrence County, New York, and southward in Pennsylvania, north and south of Philadelphia. These may represent extensions of the Franklin marble, cut off from the main body by geological events. In any case, they contain similar mineralization.

Figure 5-2. The Lime Crest quarry at Sparta, New Jersey, showing the large pegmatite body in the Franklin marble toward the upper left. A number of fluorescent minerals are found in the pegmatite, in the Franklin marble, and in the transition zone between.

The fluorescents which may be found in the Franklin marble include:

apatite	norbergite and chondrodite
calcite	phlogopite
corundum	scapolite
diopside	titanite (sphene)
edenite	tourmaline
fluoborite	tremolite
margarite	zircon
microcline	

Many of these are attractive fluorescents. Particularly attractive is the yellow fluorescing norbergite, sky blue fluorescing diopside, red fluorescing corundum, violet–blue fluorescing microcline, and yellow, pink, or orange fluorescing scapolite. Fluoborite is a rare fluorescent from these limestones.

FRANKLIN AND OGDENSBURG, NEW JERSEY

At the main entrances to the town of Franklin, New Jersey, signs are posted which proclaim Franklin as the fluorescent mineral capital of the world. Over 56 fluorescent species are recognized from Franklin and the Sterling mine. Most of these minerals do not merely fluoresce: they fluoresce brilliantly. Most are found not in thin and reluctant fluorescent coatings or small and obscure crystals, but in large, solidly fluorescent masses, and they are found in a bewildering series of startling fluorescent combinations with one another. The claim by Franklin to be the fluorescent mineral capital of the world is beyond dispute.

The two mines, one in Franklin and the other the Sterling mine at Ogdensburg about two miles south of Franklin, worked similar ores for their zinc content. Manganese was a by-product. At both locations, the ores were located within the white crystalline calcite known as the Franklin marble. The ore bodies, each thin ribbons of mineralization up to tens of feet thick, had a horizontal length of about one quarter of a mile for the Sterling Hill deposit and about twice that for the larger Franklin deposit.

The mine at Franklin ceased operation in 1954 when ore was exhausted, but the Sterling mine, despite its earlier first exploitation and smaller ore body, is still in operation. At both sites, the primary ore minerals were few in number – willemite and franklinite, and

Figure 5-3. The Parker shaft and mill at Franklin, New Jersey, around 1896. Many of the most spectacular fluorescent minerals in the world were taken out through this shaft in the years following its opening. The headframe of the shaft is visible behind the smoke stacks. (Photo courtesy of Special Collections, Sussex County New Jersey Library)

Figure 5-4. With justifiable pride, Franklin posts its claim. This claim must include the minerals of the Sterling mine at Ogdensburg, a few miles to the south.

zincite and sphalerite to a lesser extent – but the total number of minerals identified is now over 240. Many of these are unique to these two locations while others are known at only a few other places in the world. New minerals continue to be discovered in the still active Sterling mine, though now they are rare microspecimens, for the most part, rather than material in substantial masses.

However, many good, older Franklin mineral specimens are preserved in the hands of dealers or in the collections of the many serious amateurs who are found particularly within driving distance of Franklin. From time to time, some of these Franklin minerals become available on the market. They are quickly swept up by those who are able to keep tuned to the comings and goings of these specimens and who are able to produce whatever it takes to part the seller from a valued specimen.

Because of the proximity of the ore deposit to centers of population and because of its unique nature, attention was given early to the theory of the origin of this deposit. The serious study of the mineralogy of the Franklin ore deposit began at least as early as 1810, and there has been an almost uninterrupted flow of technical papers on the subject from 1819 to the present. Palache, in his famous paper on the minerals of Franklin, describes the theories in vogue at various times to explain the Franklin deposits. None appears to be sufficiently satisfactory to explain all of the facts.

A recent theory of formation was articulated by Clifford Frondel of Harvard University and John Baum, geologist at Franklin until his retirement. The theory takes note of the fact that the ore deposit resembles in its mineralogy that of metamorphosed manganese deposits and that the resemblance would have been much closer if zinc had been absent from the Franklin ore before metamorphosis. It takes note also of the metal ores now being formed in the Red Sea. From the point of view of plate tectonics, the Red Sea is considered to be a new ocean resulting from a rift and separation of Africa from Asia beginning some 20 million years ago. At the center of the rift, beneath the water surface, a spreading center exists much like that in the mid-Atlantic which was the source of spreading from which the Atlantic originated. Like the Atlantic rift, the Red Sea rift generates metal rich brines. In effect, these are hydrothermal solutions venting into a narrow and confined sea. The metals carried up in solution are oxidized by oxygen or carbonic acid in the sea and converted to insoluble fine particles which rain down and accumulate as sediments on the narrow sea bottom. The composition of these presently forming sediments casts an interesting light on the problem of the origin of the Franklin ore. One report describes the composition of the Red Sea sediments as follows:

> One class consists of materials that have entered the deep from the outside: quartz, feldspar and clay, for example, that are brought to the sea in runoff from the land, and the calcarceous debris of dead marine organisms. The other class is generated entirely within the brines themselves. It consists of various metal compounds: sulfides, sulfates, carbonates, silicates and oxides, principally of iron, manganese, zinc and copper.*

A better description of the starting point of the Franklin ores could perhaps not be made.

If one returns now to the theory of Frondel and Baum, it would picture the slow deposition of limestone, later to be the Franklin

*Degens, E. T. and Ross, D. A., The Red Sea hot brines. *Scientific American,* April 1970.

marble, as "calcareous debris of dead marine organisms." At some time after a substantial bed of limestone had formed on the sea bottom, submerged hot springs vented metals into the sea which accumulated as sediments composed of "sulfides, sulfates, carbonates, silicates, and oxides, principally of iron, manganese, zinc . . ." which would be interbedded with layers more or less rich in "quartz, feldspar and clay . . .," much as now taking place in the Red Sea. This process may have continued for many thousands of years. After some time, the sedimentation of metals ceased and the deposit was covered by more limestone.

Later, all of these submerged strata were elevated above the sea and some 800 to 900 million years ago, the limestone was metamorphosed into the Franklin marble. The metal sediment layers were caught up with the limestone and underwent similar intense heating and pressure. Possibly, the drifting apart of continents which opened this narrow sea reversed itself at that time, so that this elevation and intense pressure resulted from collision of one continent with another. This pressure transformed the minerals of the sediments into many of the numerous and complex minerals of the Franklin and Sterling mines. Where metals were concentrated and sands and clays were sparse, magnetite, franklinite, willemite, esperite, and hardystonite formed in profusion. Where silica was more abundant, rhodonite, bustamite, wollastonite, barium feldspars, scapolite, and andradite formed. The intense folding which accompanied formation of these minerals was also probably responsible for deforming and scrambling what was originally a stratified, layered deposition of minerals.

At some later time, stresses in the earth faulted the ore body. Pegmatites intruded the ore following the path of faults, and heated water ascended the faults reworking the earlier minerals into axinite, datolite, fluorite, and other vein minerals. When these events were completed, the separate ore bodies now at Ogdensburg and Franklin were probably part of one long, twisted and folded band of mineralization. Erosion removed the band of ore connecting these locations, leaving the remaining ore as separated islands in a sea of white Franklin marble, and also leaving portions of the ore exposed at the surface. These surface exposures were discovered and worked as early as the eighteenth century.

Minerals from both Franklin and the Sterling mine are usually referred to as "Franklin" minerals in recognition of the more famous of the two sites. The list of fluorescent Franklin minerals is a long one. In addition to minerals which are associated with the ore, minerals of the Franklin marble are also considered to be "Franklin" minerals when they are found with an ore bearing specimen or if

Figure 5-5. At the Buckwheat dump next to the Franklin Museum, interesting fluorescent minerals may still be found, though the dump is now reduced in size as a result of years of collecting by visitors to this famous locality.

they are found on the mine dumps. The list of fluorescent Franklin minerals includes the following:

albite	esperite
apatite	ettringite
aragonite	fluoborite
axinite	fluorite
barite	gypsum
barylite	hardystonite
bustamite	hedyphane
cahnite	hemimorphite
calcite	hodgkinsonite
celestite	hyalophane
cerussite	hydrozincite
clinohedrite	johnbaumite
corundum	margarite
diopside	margarosanite
dypingite	microcline
edenite	monohydrocalcite
epsomite	norbergite and chondrodite

- pectolite
- phlogopite
- picropharmacolite
- powellite
- prehnite
- roeblingite
- scapolite
- scheelite
- smithsonite
- sphalerite
- svabite
- talc
- thomsonite
- tilasite
- tourmaline
- tremolite
- willemite
- wollastonite
- xonotlite
- zincite
- zircon

It is difficult to nominate the minerals on this list which are particularly outstanding as fluorescents when so many are outstanding, but any of the following are candidates, particularly in combination: axinite, calcite, clinohedrite, esperite, hardystonite, margarosanite, sphalerite, willemite, and wollastonite.

Why do Franklin minerals fluoresce, and why in such a profusion of colors? It is difficult to provide a general answer, but the following possibilities would need to be considered. First, the great number of minerals available means that at least a moderate number of fluorescent species should be found if only on a chance basis. Yet it is clear that the great number of fluorescents found at Franklin – close to 20% of the Franklin species fluoresce – calls for a more fundamental explanation.

The minerals of Franklin, quite unlike those to be found in other metal ore deposits, are made up largely of oxygen dominated structures – carbonates, silicates, borates, phosphates, and the like. These are linked to the metals calcium, lead, and zinc in many cases. It is precisely such structures and chemical compositions which, on the basis of experience with synthetic fluorescent minerals, are known to produce the most vigorous fluorescents. Sphalerite, among all the sulfides which make up most metallic ore deposits, is the only one capable of fluorescence. Yet at Franklin and Sterling Hill, it is the most common sulfide present. Then also, manganese and lead are disseminated in small quantities throughout the various minerals of the Franklin deposits. Lead and manganese, together or separately, are well-recognized, powerful activators of fluorescence in many diverse mineral structures. Finally, the well-known quenchers of fluorescence – nickel, cobalt, iron – are either absent or sequestered in only certain minerals of these deposits.

A scientist, wishing to invent from his knowledge and imagination an ideal mineral deposit in which to find numerous, diverse, and brilliant fluorescent minerals, could not do better than to describe the Franklin deposit.

THE BALMAT-EDWARDS DISTRICT, NEW YORK

To the northwest of the Adirondack Mountains in New York State, there is a region made up primarily of ancient marbles, dolomites, and gneisses. This region, extending into Canada, is famous for the skarn and other mineral specimens which have been found there. The marbles of this region are much like the Franklin marbles of New Jersey and provide similar fluorescent minerals. The similarity may not be coincidental. These rock formations on the northwest edge of the Adirondacks are known as the Grenville, and many geologists believe that the Franklin marble formations are part of the Grenville.

What this means is that the rocks of the two regions probably formed under similar conditions and in the same remote period of time beginning over one billion years ago. It is even possible that the Franklin marble and the marble of the Adirondacks and nearby Canada were once one body, but this is difficult to prove since subsequent periods of folding and earth movement have disrupted any continuity of strata which may have existed. In any case, similar minerals are found, including similar fluorescents: scapolite, phlogopite, apatite, microcline, and norbergite. Undoubtedly all of the other fluorescent minerals of the Franklin marble may be found somewhere in the marbles of the northwest Adirondacks.

In one area of St. Lawrence County, New York, there is a particularly interesting band of mineralization within the marble. This consists of a narrow belt of tremolite, talc, and occasionally sphalerite found along a line extending from Balmat, New York, northeastward to Edwards, a distance of over ten miles. All three of these minerals are commercially valuable, and operating and dormant mines exist at various places along this line.

The tremolite and the talc are found in parallel beds and layers, and are mined together. At Balmat, the Gouverneur Talc Company operates both underground workings and a large open pit, recovering talc and tremolite. These are used as fillers in paints, putties and numerous other applications. The St. Joe Minerals Company mines the sphalerite. The headframe and offices of the talc company are almost within a stone's throw of the headframes and offices of the St. Joe mines at Balmat, so close are their underground operations. As shafts and stopes in the two operations pursue the ores which lie so close to one another, ample opportunity exists for one working to inadvertently break into another. This is prevented by good survey work and the use of one grid system by both companies. Even then, the drill holes of one mine have occasionally breached through to the other, and miners were able to hear one another through the wall between the mines.

Figure 5-6. A work break in a tremolite-talc mine. At various locations in the tremolite-talc belt near Balmat, New York, the ultraviolet light will show the brilliant red fluorescence of tirodite, the orange of tremolite, or the yellow of talc. (Photo courtesy of Fred Totten and Gouverneur Talc Company)

Elsewhere along this ten-mile mineralization, a sphalerite mine is found at Talcville, where talc-tremolite mining was also carried out in the past. Zinc mining too was carried out at Edwards, New York.

The origin of all of these mineral deposits is somewhat problematical as is typical of deposits so ancient. One or more periods of metamorphism at a time about one million years ago stressed, folded, overturned, and ruptured strata over the entire region. Under metamorphic conditions, dolomite may have been robbed of its magnesium which combined with silica and amply available calcium to form the large, dense beds of tremolite. Continuing metamorphism converted some of the tremolite to anthophyllite, and presumably to tirodite in regions where manganese was available. The introduction of water facilitated further changes to talc.

The lead minerals were probably deposited initially as sphalerite sediment in a shallow sea, possibly interbedding with the dolomite. Like the dolomite and associated deposits of silica, the sphalerite underwent intense metamorphism and later may have been squeezed into favorable locations within the intensely folded strata. It is possible to see in this geological history a process of formation similar to

Figure 5-7. Headframe of the St. Joe No. 3 at Balmat, New York. Over the years, a number of interesting fluorescent minerals have been found here in the course of mining zinc ore.

that of the Franklin ore body, except that the metal minerals deposited in what is now the Balmat region were less varied in the metals contained and less subject to oxidizing conditions than the sediments in what is now the Franklin region.

In recent times, the exposure of the sphalerite at the surface has allowed oxidation to take place. Some of the surface sphalerite was converted by oxidation to easily soluble zinc sulfate which trickled down into the ore body. Here, it oxidized other sulfides and was in turn reconverted to sphalerite. Thus, some sphalerite in these mines is supergene in its present origin. As far as fluorescence is concerned, this is fortunate. The great bulk of sphalerite in this district contains substantial iron, an enemy of fluorescence. However, the supergene material has a much lower iron content, and much of it fluoresces. Such sphalerite is sometimes accompanied by fluorescent secondary willemite.

Numerous other fluorescent minerals occur among those brought to the surface in the course of mining for sphalerite, or talc and tremolite. Some of these inevitably are those native to the marbles which interbed with the ore bearing layers. The fluorescent minerals include:

albite	scapolite
anthophyllite	sphalerite
calcite	talc
hydrozincite	tirodite
paragasite	tremolite
phlogopite	willemite

There are a number of interesting or outstanding fluorescents among these. Calcite has been found in translucent crystals of unbelievable size (over one foot) in the course of zinc mining. These fluoresce weak pink. Small clear crystals of sphalerite have been found, fluorescing rich yellow-orange with an almost red phosphorescence. The massive supergene sphalerite fluoresces orange under long wave. At the edge of these sphalerite masses, blue, blue-green, and green fluorescing sphalerite can be seen rarely. This is sometimes accompanied by green fluorescing willemite. Finally, scapolite has been found at the zinc mine at Edwards which duplicates in every way the intense yellow-orange fluorescence of wernerite from Canada.

In the talc-tremolite family, large tirodite crystals are found in the Talcville area with a brilliant red fluorescence, particularly strong under long wave. This may be found with fluorescing talc, tremolite, or anthophyllite to make fine color combination specimens. Tremolite and anthophyllite are found in large blocks of bladed crystals, and the size of these specimens contributes to the attractiveness of the fluorescence.

THE COLORADO PLATEAU

There is a place in the southwestern United States where four states meet. One can visit this location and stand simultaneously in Utah, Colorado, New Mexico, and Arizona. From this location, it is possible to look into any of these four states. This is the center of a geological region known as the Colorado Plateau, an elevated terrain composed of water-cut ancient sedimentary rocks. An examination of a map of the metal deposits of the West will show that this plateau region is surrounded by locations mined at one time or another for gold, silver, copper, zinc, or lead. However, the Colorado Plateau itself is barren of such metal riches.

The Spanish adventurers, departing from the city of Mexico, did not know this when they directed their march to the plateau in search of the rumored Seven Cities of Cibola. These cities were said to be rich with gold, with a treasure for the taking, richer than that taken by Cortez in the conquest of Mexico. What the Spanish invaders found were several busy Indian towns, already ancient when the Spaniards arrived. These included the pueblos of Zuni, Acoma, and others in New Mexico which were even then rich in tradition and history but not in gold.

While it does not contain precious metals, the plateau does contain a mineral treasure of the present day – uranium – in sufficient quantity that it is the major source of this valuable element in the United States. Many of these deposits were first found during the uranium boom of the 1950s, when a bounty offered by the Atomic Energy Commission stimulated a heated search by both amateur prospectors and professionals. Many means were used to support the difficult search. Geiger counters were carried by hand, and the more sensitive scintillation counters were used in aerial sweeps. Ultraviolet lights saw a second boom of utilization – the first being the search for scheelite in the early 1940s. These lights revealed the traces of fluorescent secondary uranium mineralization which sometimes proves to be an indicator of a major deposit.

Even botany was employed. "Locoweed" sometimes indicated the presence of uranium, since the plant favors ground rich in selenium which often accompanies uranium. Among the uranium deposits discovered, some of the richest are to be found in New Mexico close to the Zuni and Acoma pueblos, in the Grants region of New Mexico.

The Grants region contains the Ambrosia Lake district in McKinley County, north of the town of Grants, and the Laguna district east of Grants in Valencia County. Each of these districts contains numerous mines, open pit workings, and exploratory pits in the uranium bearing strata. These are of interest to the fluorescent mineral collector since the mines, particularly those of the Ambrosia Lake district, have furnished some outstanding fluorescent specimens of various uranium minerals.

The geology of these deposits is quite different from the Marysvale type of uranium deposit discussed previously. The uranium deposits of the Grants region are disseminated in sedimentary rock, primarily sandstone, with lesser amounts in limestone and shale. The uranium, primarily in the form of coffinite and uraninite, forms a film around the grains of the sandstone and serves to cement the sandstone. Ore tenor is about one-third of one percent or less, a low figure, but not particularly lower than the copper concentration found in disseminated copper deposits. As in the recovery of copper, uranium may be recovered by chemical leaching.

The origin of these ore deposits is presently only incompletely understood, and the age of formation is uncertain. The source of the initial uranium is obscure, but it may have been leached from volcanic ashes and tufts in the sandstone, or from earlier granitic rocks, thus pointing to an ultimate igneous source, remote in time. In the most commonly accepted view, the uranium is believed to have reached its present location and concentration through the agency of moving water. These waters moved through the porous sandstone, dissolving uranium present throughout the sandstone in small amounts. Such moving waters are an effective solvent of uranium minerals if the waters are slightly alkaline and charged with dissolved carbon dioxide, as they often are. The resulting dilute solutions of uranium carbonate moved slowly down or sideways until a bed of organic matter was reached. Typically, this would be a bed of peat or low coal. Here, accumulated hydrogen sulfide acted as an effective reducing agent. When the dilute uranium bearing solution encountered this region, the uranium salts were converted to an insoluble form and deposited. Over the course of time, thick deposits of the uranium minerals were built up. As evidence of this action, the uranium deposits are often found intimately intermixed with organic matter, and carbonized leaves and twigs are sometimes encountered while mining. Sometimes, twigs or logs completely replaced by uranium minerals are also found.

The deposition of these ores appears to have taken place at different times over a vast epoch. The earliest may date to somewhat over 100 million years ago, and deposits may still be forming. Rather, the deposits may still be moving, for there is evidence that the process of solution and redeposition is still taking place, with an earlier deposit being stripped by slowly moving underground waters which move the location of the concentration slowly forward.

In the mines and pits where the uranium ores are found, faults, fissures, and mine walls may be covered with effervescences of secondary uranium minerals. Numerous specimens of outstanding fluorescent zippeite and other fluorescent minerals from these deposits are now in many collections.

The fluorescent minerals found in the Grants region include the following:

andersonite
autunite and meta-autunite
bayleyite
becquerelite
phosphuranylite
schroeckingerite
uranopilite
zippeite

Generally similar fluorescent minerals should be expected from the White Canyon and San Rafael Swell areas in Utah, Monument Valley in Arizona, Paradox Valley in Colorado, and outside of the Colorado Plateau in deposits in Wyoming. Most of the fluorescents listed above can be considered as scarce to rare.

EVAPORITE MINERALS

Dry lakes of the deserts of the West, playas as they are called, are intermittently water filled after a desert rain. Once wetted, they may dry rapidly, particularly in summertime due to the high heat and low humidity which generally prevails. After the water evaporates, a white film of mineral deposit is left. In some of these lakes, the accumulation of evaporite minerals over the span of thousands of years has produced thick mineral beds. Some of these lake deposits are worked commercially for their mineral content. Searles Lake at Trona, California, is an outstanding example of such a commercially developed evaporite mineral deposit.

Searles Lake, situated in the northern Mojave Desert, is about five miles wide from east to west and about seven miles long from north to south. Seen from California Highway 178 which runs along the western side of the lake, Searles appears to be a vast expanse of white stretching to the foot of the mountains in the far distance. On the lake, other details are evident. The hard white surface is crisscrossed by elevated roads used by the Kerr-McGee Chemical Corporation in the course of its work. There are natural ponds at several places on the surface and several artificial solar ponds as well, used for the concentration of brines during the hot summers.

The lake is composed of layers of evaporite minerals or salts, alternating with layers of mud, to a depth of about 700 feet. The salt layers are generally porous and brine filled, with composition differing in different layers and from the center of any layer to the edge. Minerals are extracted for commercial purposes by pumping out brines, and the different composition of the brine at various depths allows the company to select brines whose composition is best suited for chemical production.

Once a year the company opens the gate to mineral collectors. To prepare a worthwhile harvest, the company may drill down into the interesting layer called the Upper Salt or the Upper Structure, open a cavity at the bottom which quickly fills with crystals suspended in brine, and blast these crystals to the surface with an air hose. The resulting huge pile of crystals may include halite, hanksite, trona, and possibly gaylussite and pirssonite loosened from the mud layer below the Upper Structure.

These and the other evaporite minerals found at other depths are of interest to the fluorescent collector since almost all of these mineral species will fluoresce and phosphoresce, at least sometimes.

Figure 5-8. Collecting day at Searles Lake, Trona, California. Crystals of a number of minerals for which Searles Lake is known are gathered by collectors. On the far edge of the lake on the left, part of the Kerr-McGee processing facility can be seen. (Photo by Steve Mulqueen)

The Searles Lake salts began to be deposited perhaps one-half million to one million years ago. During periods of extreme melting of the glaciers in the Sierra, water was plentiful in the basin valleys. During such times, a series of interconnected lakes existed along a north-south line. The northernmost was Mono Lake whose waters drained southward into Long Valley and then into a much enlarged Owens Lake. Owens Lake fed water southward into a then water-filled China Lake. In turn, China Lake coursed into Searles along what is now the location of the highway connecting Ridgecrest and Trona, California.

During such wet periods, Searles was water filled, sometimes to a depth of almost 700 feet. Inflowing waters carried sediment in suspension and minerals in solution. The sediments dropped to the lake bottom, forming the mud strata now found at various depths in the lake. During the dry periods, evaporation of the waters precipitated the minerals to form the salt beds. Alternating periods of plentiful and scarce water account for the alternate layers of salts and mud now found in the course of drilling in the lake.

Certain of the elements of which the many salts of Searles Lake are formed, sodium and calcium in particular, can be explained as decomposition products of the feldspars of the Sierra. Certain other elements, notably chlorine, sulfur, boron, and tungsten – the lake salts contain an immense amount of disseminated tungsten – may require a different explanation. These are typical constituents of hydrothermal fluids, and it has been suggested that such elements in the salts of Searles originated from a hydrothermal source. In this view, the sources of these elements are thermal springs in the Long Valley area, which contains hot springs and geysers up to present times.

Searles Lake produces perhaps the greatest diversity of minerals among the various evaporite mineral deposits. These include the following fluorescents:

aphthitalite (glaserite)	northupite
borax	pirssonite
burkeite	sulfohalite
halite	thenardite
hanksite	trona
nahcolite	tychite

The mineral gaylussite is not included because specimens from Searles which have been examined fluoresce only due to mud inclusions. Gaylussite from an African source fluoresces strongly, however, so that it is possible that fluorescing gaylussite will be found at Searles. Fluorescing searlesite is known from the Green River, Wyoming, saline deposit and thus may be found to be a fluorescent at Searles.

Surface or buried evaporite mineral deposits are found in other parts of the country including Nevada, Wyoming, New Mexico, and New York. These may produce some of the same fluorescent minerals. The Green River deposits mentioned above, located in southwestern Wyoming, produce some fluorescent minerals identical to those of Searles, and at least one fluorescent mineral – shortite – peculiar to that location. Finally, the borate deposits discussed next are also evaporite deposits, treated separately because of the distinctive minerals for which they are noted.

BORATE DEPOSITS

To the east of Searles Lake is the lowest dry land in the Western Hemisphere – 283 feet below sea level – near a spot called Badwater in Death Valley. This deep valley acts as a natural sink for waters draining downward onto the valley floor. Since such waters contain minerals in solution, and since the extraordinary summer temperatures of Death Valley promote rapid evaporation, mineral deposits are formed in the lowest portions of the valley.

The valley floor was an early and important location for exploitable borate minerals. Balls of ulexite were discovered here on the dry surface in 1881, and a small processing plant, the Harmony Borax Works, was set up in the valley to produce borax from the ulexite. The process required the reaction of ulexite with soda ash in boiling water. Cooling then crystallized the borax, but summer temperatures were so high that the hot solutions refused to cool. Then, another deposit was found some distance southeastward, at Amargosa. There, summer temperatures were cooler, only 110°F on the average. Processing in the summer was carried out there and in the winter at the Harmony Works. The famous twenty mule teams were employed in carrying 37-ton loads of borax from Harmony to the nearest rail head at Mojave, some 165 miles distant.

A few years later, high up in the Black Mountains which form the eastern wall of Death Valley, borates were found in thick layers bedded within the numerous rocky strata which form these mountains. Valley workings were abandoned in favor of the richer mountain deposits which are still being mined today.

The Black Mountain deposits were originally lake evaporite deposits. These now elevated mineral formations were set down some ten million years ago when the terrain in which they are now found was substantially lower. Afterwards, great blocks of rocky strata bearing these borate deposits faulted and tilted upward, thus elevating the borates. At the same time, the block of the earth's crust, upon which the valley now rests, dropped down between the tilting formations to the east and west. The area is apparently still in motion, geologically speaking, and the tilting upward of the great blocks of rocky strata which form the mountains to either side of the present valley may still be taking place. If so, the mountainside deposits are still increasing in elevation while the valley floor may be sinking lower yet.

Volcanic activity was prominent in earlier times in the area. Hot springs associated with volcanic activity were the likely source of the boron which is an essential constituent of the borate minerals. The discharges probably contained ulexite in solution. In time, beds of ulexite were deposited which subsequently became buried. As layer upon layer of overburden accumulated, pressure and heat converted some of the ulexite to probertite. Then, over the further course of time, calcium bearing waters percolated down to these ulexite-probertite strata which furthered the enrichment of calcium in these minerals, converting them to inyoite, meyerhofferite, and colemanite. These deposits remained buried while the overburden became consolidated into the rocky strata, which are now the Black Mountains.

A little over 100 miles southwest of Death Valley, in the middle of a portion of the Mojave Desert which seems as featureless as Death Valley is starkly featured, one of the major borate deposits of the

Figure 5-9. Part of the vast subsurface borate deposit at Boron, California, is visible in this photograph. A number of fluorescent borates have been found here. Interesting fluorescent borate minerals are also found in the deposits in the mountains which form the east wall of Death Valley. (Photo courtesy of Jim Minette and U.S. Borax)

world is mined in a large open pit operation at Boron, California. This borate deposit was discovered well after those of Death Valley, despite the fact that Boron is substantially nearer to the populated Pacific coast. The reason is that balls of ulexite lie visibly on the surface of Death Valley and borate bearing strata are exposed in the adjacent mountains. However, it was only after a resident of the Mojave attempted to drill for water at what is now Boron that the deeply buried borate beds were discovered. The origin of the deposit at Boron can be taken to be similar to that of the other deposits – a lake into which mineral rich springs drained, followed by burial in sediments as the sources of ample water disappeared. Here, however, sodium bearing borates – borax and kernite – remain predominant, indicating that the influence of calcium bearing waters was less significant at this location. Here the calcium borates occur as a thin shell of altered minerals around a huge lens of borax and kernite. Here also, earth movement has not thrust up the deposits.

Many of the borates from both Boron and Death Valley are identical, and many of them fluoresce. These may include:

colemanite
gowerite
hydroboracite
meyerhofferite
priceite
probertite
tincalconite
tunellite
ulexite

Of these minerals, colemanite can be considered the outstanding fluorescent because of the frequency with which specimens may be found that fluoresce, high brightness, good phosphorescence, and the fact that handsome crystal specimens are obtainable. At Boron, arrowhead shaped crystals of colemanite that fluoresce pale white have been found upon tan balls of calcite that fluoresce yellow, thus making a particularly attractive fluorescent. Other borate minerals are quite scarce or, like gowerite, rare.

While Boron and Death Valley are the most prominent borate sites, numerous other locations exist which may provide fluorescent specimens, particularly in California.

6
Fluorescent Minerals—Description

The description of fluorescent minerals in the first part of this chapter is limited to those found in the United States. This is a necessary limitation since it is not possible to examine and report in detail on all of the fluorescent minerals of the world. A second and shorter part of this chapter describes a number of fluorescent minerals worldwide where these differ from those of the United States.

To describe the fluorescents of the United States and the sites from which they come is itself a major challenge. There are, for example, over 1,000 listed sources of scheelite in this country when active and abandoned mines, old dumps, pits, and exploratory prospects are considered. Locations for the more commonly found fluorescent minerals are too widespread and too numerous to be completely covered here. For the rare minerals, a different problem exists. If the material is known from only one or a few places, no great problem is presented in the identification of these localities. However, the fluorescent properties of some rare minerals may not have been described in the literature and many were not accessible for examination. Such minerals may not be described here. Also, some common minerals may fluoresce only very rarely or, if ordinarily fluorescent, may in some rare instances fluoresce in some very unusual way. These exotic responses may be unrecognized, or known only to a few, and may be slighted here.

Even if these problems did not exist, it is not possible to complete a description of fluorescent minerals. New minerals are being discovered at the rate of several dozen per year, and since a certain number of these will be fluorescent, the problem of describing the fluorescence of minerals is a continuing one.

Within these limits, a diligent effort has been made to cover the fluorescent minerals of this country. Many spectacular fluorescent minerals or astonishing combinations of various fluorescents can be found here and, to the extent possible, these are described. There are also many unspectacular fluorescent minerals, and these are also described since a scientific or methodical collection will not be limited to the impressive pieces only. As the collector advances in sophistication, these less spectacular, often rarer, specimens will be sought and will enter the collection.

IMPORTANCE OF OBSERVATION

The fluorescent response of minerals has been known for some time, and systematic lists and descriptions have been made since at least the beginning of this century. Some of these lists have been based on careful, direct observation of the fluorescence under whatever source of ultraviolet excitation was available at that time. Often, the descriptions which have been provided in this way do not match the response seen under modern ultraviolet lights, because either no filters or only poor filters were available and visible light leakage masked or distorted the color response. Also, weak ultraviolet sources may not have revealed the fluorescence of the mineral at hand. As a result, some early lists can mislead the present day collector. Other, more recent, lists and descriptions have been compiled from secondhand sources, and errors have crept in by quoting from others.

As a consequence, many presently available descriptions of fluorescent minerals contain frequent and annoying inaccuracies. Nothing substitutes for direct, firsthand observation when preparing a description of mineral fluorescence. That view has been followed in this book. Wherever possible, direct observation and examination of the mineral under short and long wave ultraviolet have been carried out by the writer, using modern equipment, and the descriptions in this book are based primarily on these observations. In a few cases, particularly where a few rare minerals are concerned, this has not been possible, and the observations of others have been used. While such observations and descriptions may be valid, one cannot always be certain. For this reason, descriptions of fluorescence taken from elsewhere are clearly noted. Whenever a fluorescent response in this book is described as "reported," as in "the fluorescent response of this mineral from Joplin is reported to be gray-white...," then it is based on an outside reference. In all other cases, the description is based on the author's own observations of the fluorescence of the mineral in question.

ORGANIZATION

In the past, descriptions of fluorescent minerals have been organized alphabetically or, in some cases, according to the fluorescent color. Both approaches have practical advantages, but an organization of fluorescent minerals according to the underlying fluorescent properties would be far better. Thus, it would be desirable to group fluorescents according to an activator they have in common or according to the electron energy levels which participate in the fluorescence producing energy exchange. Such an approach would furnish a scientific and systematic organization of fluorescents. Unfortunately, far too little in known at present about the activators and energy levels specifically responsible for the fluorescence in minerals. That at least some groups exist showing commonality in fluorescence activator is evident. Manganese is widespread as an activator, frequently substituting for calcium, and commonly produces yellow, orange, or red fluorescence response. Minerals activated by manganese in place of calcium could be organized as a fluorescent group, and when manganese produces a similar response in closely related minerals such as pectolite, wollastonite, and miserite, the grouping can be made with a high degree of confidence.

In addition to the pectolite group, another grouping of related minerals shows an apparent common fluorescence relationship. Thus, celestite, anglesite, and barite are often fluorescent. Aragonite, cerussite, strontianite, and witherite are also frequently fluorescent, usually in a blue-white or yellow-white color. In addition, a large number of hydrous borates are fluorescent in either a blue-white or yellow-white color. Within each of the groups one may suspect a common activator or fluorescent system.

There are other possible fluorescent groupings which are contrary to a Dana system of classification. Thus, the lead minerals anglesite, cerussite, leadhillite, phosgenite, and matlockite display a remarkably similar yellow-orange fluorescence, which is particularly strong under long wave ultraviolet; this suggests that a common activator is at work, having more to do with the metal lead present in each, than with the nonmetallic component which distinguishes these minerals.

However, the vast majority of fluorescent minerals do not fit neatly into any evident groupings based on presently understood underlying commonality, so that it is not possible to fully organize fluorescent minerals according to a fixed scientific system. For lack of another valid approach, and rather than using an alphabetic or color basis of organization, the minerals described here are listed according to the Dana system. This has the advantage that at least some fluorescents fall into groups under this system, and other groupings may become evident on the basis of further investigation. It has the additional advantage of being familiar and comfortable to most collectors.

Only one exception is made. These are the green fluorescing uranium minerals. These are not grouped together in the Dana system but certainly deserve to be treated together as a fluorescent group. Their fluorescent color response is essentially indistinguishable one from another, and it is known that this fluorescence is due to a common component, the "uranyl group," built into the chemistry of each such mineral; these minerals are described as a separate group in this book.

DESCRIPTION OF FLUORESCENCE

The fluorescent responses described here can be seen under hand held ultraviolet lights with tubes and filters in good condition. Higher intensity lights may produce other responses and particularly may show fluorescences which are too weak to be seen under the hand held light. However, since the majority of collectors will have hand held lights, a limitation to what such lights will reveal is called for.

By no means do all specimens of a given mineral fluoresce. In some cases, scheelite for example, fluorescence may seem to be so common that it might appear to be a universal property of scheelite. In other minerals, the occurrence of the fluorescent responses described here may be so uncommon that diligent search over many specimens of the mineral species is required before a suitable fluorescent sample is found. These limits should be kept in mind when comparing these descriptions with any specimen in hand.

Part A: Fluorescent Minerals of the United States

SULFIDES

Sphalerite, (Zn, Fe)S.

Possibly the most interesting and consistently fluorescing sphalerites are found at Franklin, New Jersey, and the Sterling mine at Ogdensburg, New Jersey. These sphalerites occur as resinous blebs or masses, colored gray, yellow-gray, tan, or pink-tan in ordinary light. In contrast to most of the other numerous and famous fluorescent minerals from these two mines, sphalerite fluoresces best under long wave ultraviolet. Material from Franklin usually fluoresces bright, clear orange with an enduring orange phosphorescence. The fluorescent response of the material from Ogdensburg is more varied. It may be an orange with a slight brown tone, pink-orange, brick red, or orange-yellow. Frequently, the phosphorescence is of the same color. This sphalerite is found with dark green fluorescing willemite, or pink or red fluorescing calcite, and occasionally with a violet fluorescing calcite, all under long wave. Such combinations are extremely attractive.

In addition to these orange fluorescences, many specimens also show a bright sky blue fluorescence under long wave. This response is seen usually surrounding the orange, at the boundary between the sphalerite mass and enclosing minerals. Rarely, the entire sphalerite mass will fluoresce this intense, sky blue with little or no sign of an orange response. This blue response often carries forward into a phosphorescence of the same color. A "nonlinear" fluorescence effect is evident in such orange and blue fluorescing specimens, meaning that as ultraviolet intensity is increased, the brightness of the orange response does not increase as rapidly as the blue. Thus, when the ultraviolet light is held at some distance from the material, the orange response will dominate, but as the light is brought closer, as the intensity of the ultraviolet illumination is thus increased, the blue response will begin to assert itself and will at last be the dominant response.

Intense orange fluorescing sphalerite is found at the St. Joe No. 3 mine at Balmat, New York, sometimes accompanied by green fluorescing willemite. Both minerals are supergene or secondary in formation. For the sphalerite in particular, this mineraligical reworking appears to be necessary in order for fluorescence to occur, the great bulk of sphalerite from this mine being dark and nonfluorescent. Very, rarely, at the boundary between this sphalerite and the host rock, an intense blue fluorescence is evident under long wave, much as in the material at Ogdensburg, New Jersey. More unusually, some of this will fluoresce blue-green and yet other portions will fluoresce green under long wave. Little green or blue response under short wave is evident. Similarly, rare shalerite material from the Old Hornsilver mine in Frisco, Utah, will fluoresce in various sections in orange, blue, blue-green, and green. If one assumes that this blue and green fluorescing material is indeed sphalerite, which appears to be the case, a very interesting light is cast on the possible activators involved, as discussed below.

However, let us return for the moment to the No. 3 mine at Balmat where another most unusual fluorescent sphalerite has been found. This consists of glass-clear crystals of a light yellow color up to approximately three quarters of an inch in size. Some of these fluoresce yellow-orange, best under long wave. The phosphorescence is an astonishing deep orange, almost red in color.

Orange fluorescing sphalerite is also found at Silverton, Colorado, with the other sulfides pyrite and galena. Sphalerite has been found as granular gray masses at the Queen mine at Bisbee, Arizona, fluorescing an astonishingly bright yellow-orange; it may be accompanied by red fluorescing calcite. Sphalerite, variously reported as fluorescing red, orange, or yellow, is found at Bonanza, Colorado; Cherokee County, Kansas; and the Banner district, Idaho. As this mineral is found in many places across the country, many other localities can be expected to produce fluorescent specimens. It should be noted that many orange fluorescing sphalerites are also triboluminescent. If the specimen is scratched or stroked in the dark, a flash of orange will be produced along the stroke.

The source of the fluorescence in sphalerite has been the subject of more study and investigation than that of any other fluorescent material, and such study has gone on for at least a century. As a result of these extended investigations, numerous activators have been identified. For example, copper, lead, tin, manganese, and even iron have been shown to be capable of producing a red-orange response in sphalerite under certain conditions, and copper, silver, and iron have been shown to produce a blue response under other conditions. Since sphalerite from Franklin and Ogdensburg is accompanied by manganese bearing minerals, we may suppose that manganese is present in this sphalerite and is the activator of the various shades of orange. It may be the presence of silver which is responsible for the blue response, and in fact, synthetic silver activated sphalerite has long been used as the blue phosphor in color TV picture tubes. Interestingly, pure sphalerite, that is, material with no activator present, can also fluoresce blue. However, the blue and green response seen in the material from the St. Joe and Old Hornsilver mines – if we assume again that it is sphalerite which is being seen – strongly suggests another cause. It is well established that copper and aluminum, acting together as co-activators, can produce a blue, blue-green, or green response in sphalerite, depending on the concentration ratio of these two elements. In some ratios, even a red response may be produced. It is thus likely that copper and aluminum (several other elements may act in place of aluminum) produce these blue and green responses, and may produce at least some of the red or orange responses also.

HALIDES

Halite, NaCl.

Halite is the mineral of our familiar table salt. Halite crystals from the Salton Sea and Amboy, California, and from Carlsbad, New Mexico, fluoresce bright red-orange under short wave ultraviolet. The same response is reported for material from Borego Valley in California and from the Petersen salt spring at Tygee Creek in Idaho.

The color of the halite fluorescence is remarkably similar to the familiar red-orange fluorescence of calcite from Franklin, New Jersey. This similarity is not completely coincidental. The activator system in the two minerals is the same. It has been known since the last century that manganese is an essential activator of red fluorescence in calcite, and manganese has been recognized as an activator in halite since 1926. The system of activators, for there are two activators at work in both calcite and halite, was not fully understood until the early 1940s. Experiments with synthetic halite into which manganese was introduced confirmed that manganese produces red luminescence when the material is excited by an electron beam. Yet this activator by itself is insufficient to produce fluorescence under ultraviolet. Further experiments showed that lead must also be present, though, as with manganese, only very small amounts are needed. Fluorescence can be produced with less than one-hundredth of one percent of manganese and a like amount of lead. Lead atoms are able to absorb short wave ultraviolet which manganese atoms are unable to do. The energy of the ultraviolet is then conveyed to the manganese atoms which emit red light.

While the same two activators are at work in both calcite and halite, it is perhaps remarkable that the fluorescent color produced is so similar in both. It would be expected that the environment seen by manganese atoms in such different minerals would cause some distinction to be produced in the color or shade of the fluorescent light. For example, the same two activators are at work in apatite, and in that mineral a yellow-orange color is produced rather than red.

While the red fluorescence of halite is the most familiar and the most spectacular, other fluorescences are known. Halite from Amboy, presumably without manganese, sometimes fluoresces pale white, and halite from Borax Lake in California is reported to fluoresce similarly. Nice specimens consisting of cubes of halite are found at Zuni Salt Lake in Catron County, New Mexico, which fluoresce white under long wave, weaker under short wave.

Fluorite, CaF_2.

Fluorite specimens in the form of groups of large cubic crystals of various colors from Cave-in-Rock and Rosiclare, Illinois, are among the most attractive to be found in the United States. As crystal specimens, they are known the world over, but they are less illustrious as fluorescents. Those crystals which are dark blue or purple appear to fluoresce rarely, if at all. Sometimes colorless or yellow crystals from these locations will fluoresce yellow, best under long wave. Often, light colored fluorites are found as phantoms within purple or blue crystals, and the yellowish fluorescent light emanating from the inner crystal will emerge as violet because it is tinted by the outer crystal. The activator for such yellowish fluorescence is said to be petroleum trapped within the fluorite. Sometimes a fleck of a tar-like material is found in the fluorescing portion of the crystal, which lends credibility to this view. These Illinois fluorites are sometimes found with other fluorescent minerals, including calcite, barite, witherite, strontianite, and benstonite.

Somewhat similar fluorite is found at Elmwood, Tennessee, as purple crystals enclosing colorless phantoms. Again, the inner crystal may fluoresce a strong yellow under short wave, and this may illuminate the purple portion of the crystal so that it appears to be a purple fluorescent. Microscopic examination reveals that the fluorescing zones of these crystals are riddled with balls and wisps of a black tar-like material, and the inference of an organic activator is now strong.

Beautiful crystals of fluorite in the form of colorless, yellow, honey colored, or brown cubes are found in limestone quarries in Ohio. The materials from Clay Center are particularly well known and much sought after. These fluoresce yellow-white, butter yellow, or lemon yellow, best under long wave, and show a particularly strong and enduring phosphorescence. Again, it has been suggested that an organic activator is at work here. Small but attractive crystals of fluorite are found at the Caldwell quarry in Danville, Kentucky. This location is not part of the Illinois-Kentucky fluorspar zone but, rather, is some distance east near Louisville. The fluorite crystals are generally small, clear, and colorless, like miniature ice cubes, or they may be faintly tinted with purple. The fluorescence is clear white, best under long wave, with phosphorescence of similar color.

These responses are interesting but are not the most typical or characteristic response of fluorite. Fluorite is the senior fluorescent mineral since the phenomenon of fluorescence was named after fluorite by George Stokes during the nineteenth century. When he held a specimen from one of the English fluorite mines in the ultraviolet portion of the spectrum, it burst forth with blue light. Stokes was fortunate in employing the fluorites of England for these experiments since these are unusually sensitive and brilliant under ultraviolet. This blue fluorescence of fluorite, more exactly a violet-blue, is particularly strong under long wave. It is characteristic of much fluorite throughout the world.

In the United States, such fluorite is found in innumerable places, usually closely accompanying metallic ores of zinc, copper, lead, or manganese. Yet rarely if ever do our American fluorites equal those of England in fluorescent brilliance.

The American Tunnel mine at Gladstone near Silverton, Colorado, yields veins of massive fluorite and groups of cubic fluorite crystals. This material is colorless or light green in daylight, and the fluorescence is bright violet-blue, approaching the English material in intensity. It is often found with red fluorescing calcite of moderate brightness. In Arizona, the Kofa Mountains near Yuma provide fluorescent veins of fluorite and calcite, making an attractive two-color blue and red fluorescent of good intensity. In New Mexico, the Royal Flush mine in Bingham produces light blue, cubic crystals of fluorite represented in many collections because of their attractiveness. These fluoresce violet-blue and are sometimes accompanied by red fluorescing barite. The nearby Blanchard mine yields a very unusual fluorite, made not of the usual cubic crystals but of large, clear, colorless balls of fluorite slightly frosted on the outside, fluorescing blue. Another unusual fluorite is found at the Monarch mine in Yavapai County, Arizona. These are crystals in octahedral rather than cubic form. In ordinary light, these crystals are chalk white and completely opaque. They also fluoresce violet-blue.

A blue or, more nearly, a violet fluorescence is found in some of the lightly colored Illinois fluorites, but this fluorescence is usually weak. Light or pale green material from West Moreland, New Hampshire, fluoresces weak blue. Deep grass green fluorite from Mesa and San Juan counties in Colorado fluoresces similarly.

The cause of the violet-blue fluorescence in fluorite is considered to be established with confidence. It is said to be due to the presence of the rare earth element europium as an activator in concentrations of one part in a thousand or less. This element is a frequent substitute for calcium in the crystal lattice of some minerals primarily because of the similar size of calcium and europium atoms. Others of the rare earth elements are also sometimes present and may modify the fluorescent response, or produce new responses, including perhaps those described below. The source of the europium may be the metal ore deposits with which blue fluorescing fluorite always seems to be associated. Blue fluorescing fluorite can be taken as an indicator of metals near by.

Among the most interesting and unusual fluorites are those which fluoresce in different colors under short and long wave ultraviolet. For example, colorless fluorite crystals from Fort Wayne, Indiana, sometimes fluoresce white under short wave. Under long wave, the fluorescence is violet. This appears to be an organic activator at work under short wave, overridden by the blue response of europium under long wave.

Fluorite from the old scheelite mine at Trumbull, Connecticut, is an outstanding two-color fluorescent. The fluorite is colorless, pink, or tan in ordinary light. Under long wave, the fluorescent response is the usual shade of fluorescent fluorite blue, but under short wave, some or all of it will fluoresce green. The portion which does not fluoresce green fluoresces the usual blue as under long wave. The portion fluorescing green under short wave will have a beautiful phosphorescence of long duration, and this green phosphorescence is produced by both short and long wave excitation. Some of the fluorite from the Sterling mine at Ogdensburg, New Jersey, produces similar responses. Like the Trumbull, Connecticut, material, these fluorites are pink or tan in daylight. Unlike the Trumbull material, the Ogdensburg fluorite fluoresces identically under short and long wave. In much of the Ogdensburg material, the fluorescence is a dull blue-gray tinged with green in places. However, when the ultraviolet light is removed, these fluorites produce a beautiful and long lasting green phosphorescence of the same kind as the Trumbull material. The green response of these fluorites – those from Trumbull, Connecticut, and Ogdensburg, New Jersey – is best seen on a freshly broken surface, in which case the green fluorescence is brilliant. Sunlight destroys the green fluorescence and phosphorescence. In fact, a freshly exposed surface of fluorite from the Sterling mine at Ogdensburg may fluoresce a brilliant green, but if the specimen is held near the light of a bulb for even a few seconds, the green fluorescent capability will be virtually destroyed. Interestingly, these fluorites will fluoresce green as a result of exposure to the light of a light bulb, though this capability too disappears quickly.

These Ogdensburg fluorites, and probably the Trumbull fluorites also, are thermoluminescent. If such fluorite is heated, by placing a piece on a heated electric grill for example, it will begin to glow an intense green or turquoise green. This beautiful light will continue for several minutes and then expire, even if heating is continued. When the fluorite has cooled, the daylight pink color will be gone. So forever will the ability to fluoresce, phosphoresce, and thermoluminesce in green. The specimen will now only be capable of a drab version of the conventional fluorite blue fluorescence. Fluorite which thermoluminesces green has been called chlorophane, and has been found also at Amelia Court House in Virginia.

The activator at work in green fluorescing fluorite is not well established, but certain rare earths, uranium, and manganese are each said to be able to produce a green response in fluorite.

A very unusual fluorescence is produced in fluorite from Rio Arriba County, New Mexico. This material is opaque, and greenish or gray in daylight. Under long wave, the fluorescence is violet. Under short wave, the fluorescence is a surprising and intense canary yellow, and both wavelengths produce a greenish phosphorescence. A fluorite, apparently yttrofluorite, from the Black Cloud mine in Park County, Colorado, fluoresces typical fluorite blue under long wave, but under short wave, it fluoresces butter yellow to orange-yellow. An identical fluorescence is seen in fluorite from the Edison mine in Sussex County, New Jersey, and is attributable to yttrium which has been found to be present.

Additional information on fluorite will be found in Part B of this chapter.

Tveitite, $Ca_{1-x}Y_xF_{2+x}$.

Tveitite is one of those exquisitely rare minerals which are mentioned only because of possible scientific interest, rather than because of any potential for collection. The mineral is a yttrian fluoride discovered in specimens from the now extinct Berringer Hill pegmatite in Llano and Burnet counties, Texas. This pegmatite is now said to be well under Lake Buchanan. The tveitite was discovered as small whitish inclusions in yttrian fluorite from the pegmatite and is best distinguished from the fluorite by fluorescence. Under short wave ultraviolet, the tveitite is reported to fluoresce yellow-orange while the fluorite fluoresces cream to pale yellow.

Calomel, Hg_2Cl_2.

A rare chloride of mercury, calomel is found at the old mercury mines at Terlingua, Brewster County, Texas. It is sometimes found as fine white coatings on a host matrix. Rarely, solid masses of this mineral are to be seen. It is sometimes found intermixed with a fine grain, yellow-green material which may be terlinguaite. Calomel fluoresces best under long wave ultraviolet. The color is a bright deep orange or, in some specimens, red tinged with orange, somewhat like the color of red-orange fluorescing calcites from Franklin, New Jersey. Calomel is also found at Redwood City, California, fluorescing bright deep orange identical to the material from Terlingua.

Research with synthetic calomel shows that the fluorescence is most strongly produced by 3400A ultraviolet, a wavelength generated in abundance by long wave lights. The visible fluorescent light produced has its highest intensity at 6200A. This wavelength is between red and reddish orange. It has been suggested that the activator of fluorescence in calomel consists of paired mercury atoms of different electrical charge. Presumably, charge transfer may be involved in producing this fluorescence.

Terlinguaite was mentioned above and has been described as being fluorescent yellow. A good number of terlinguaite specimens were examined and none were found to fluoresce. The fluorescence of terlinguaite is open to question.

Matlockite, PbFCl.

Matlockite is a quite rare mineral found at the Mammoth mine at Tiger, Arizona. This matlockite is an occasional fluorescent. An excellent crystal of matlockite from Tiger is owned by the Desert Museum near Tucson, Arizona. This crystal is white and disk shaped, approximately one inch in diameter, and fluoresces bright yellow under short and long wave ultraviolet.

OXIDES AND HYDROXIDES

Zincite, (Zn, Mn)O.

Zincite, an oxide of zinc, is almost uniquely a product of the mines at Franklin and Ogdensburg, New Jersey. Nowhere else does it occur in such large amounts and in such attractive displays of daylight color including yellow, orange, red, and red-brown. This coloration is usually said to be due to the presence of manganese in the zincite.

White zincite is produced in the flues of zinc smelters and is often outstandingly fluorescent. Natural zincite, however, is an inconsistent and not outstandingly bright fluorescent, and so the fluorescence of natural zincite was not widely known until recent years. It is also true that the most richly colored specimens are the ones most sought by collectors, and as these fluoresce least, this also may explain why the fluorescence was not discovered earlier. Red zincite does not appear to fluoresce, and orange zincite fluoresces only occasionally.

Red or orange zincite frequently is found with calcite, and while such zincite is nonfluorescent, the interface between these two minerals is sometimes filled with honey colored, resinous zincite or with a white, powdery zincite. Such zincite is frequently fluorescent under long wave. The fluorescent color seems to vary somewhat with the light source: black light sources produce a yellow response, and hand held long wave mineral lights, an attractive greenish yellow response. This response could be taken to be due to the presence of willemite, except for the lack of substantial response to short wave.

Synthetic zincite in which no impurities are present sometimes fluoresces green or yellow, and in that case it is believed that the activation is due to a surplus of zinc atoms beyond those required to form the zincite structure. This may also be the cause of fluorescence in natural zincite. One can also speculate that the red zincites do not fluoresce because of an excess of manganese which may quench fluorescence.

Brucite, $Mg(OH)_2$.

Brucite is found in veins and fissures in serpentine as thin bladed, translucent, white pearly crystals, sometimes up to several inches long. Brucite from numerous quarries in the serpentine belt on the Maryland-Pennsylvania border fluoresces blue-white or sky blue under long wave. Sometimes the response is quite bright. Phosphorescence is usually an intense and long-lived blue-white or green-white. The response under short wave is similar but weaker. Similar fluorescent brucite is seen in material from Hoboken, New Jersey, and the Tilly Foster mine in Brewster, New York.

There have been reports of red fluorescing brucite, presumably due to the presence of manganese, but the validity of the reports and the source of the material are uncertain.

Corundum (and Ruby), Al_2O_3.

Pink or red corundum is sometimes found in calcite at Franklin, Ogdensburg, Sparta, and elsewhere in the Franklin marble of New Jersey. While the daylight color of such material varies, it is rarely an intense enough red to rank as ruby. These pink or red corundums fluoresce a vivid and intense cerise or cherry red under long wave, with little or no response to short wave. Ruby from Franklin County, North Carolina, is also reported to fluoresce cerise under long wave.

The activator for such fluorescence has been established to be chromium substituting for the aluminum atoms of the mineral. Interestingly, this chromium is also partly responsible for the pink to red color of the mineral in ordinary light.

The Franklin marble in New Jersey also produces blue corundum or sapphire, as well as colorless corundum. These sometimes fluoresce in a weak pink color under long wave. The activator for this response is uncertain since a number of possible activators can produce such shades in corundum.

CARBONATES

Carbonates of Monovalent Bases

Nahcolite, $NaHCO_3$.

A white, fibrous mineral from Riffle, Colorado, nahcolite fluoresces weak cream-yellow under both short and long wave ultraviolet. Material from Searles Lake at Trona, California, is reported to fluoresce blue-white under short wave.

Trona, $Na_3(CO_3)(HCO_3) \cdot 2H_2O$.

The interesting mineral trona, sometimes white, yellow, or orange, may look much like blady ulexite but it is a carbonate rather than a borate. Trona from Westvaco, Wyoming, fluoresces white with a yellowish or bluish tone. Other specimens fluoresce yellow-orange. The response is stronger under long wave than short wave, and there is a short phosphorescence. Crisscross or jackstraw prismatic crystals of trona are frequent in certain salt layers at Searles Lake. Fluorescence is bright blue-white under long wave with some phosphorescence. The response is weaker under short wave.

Carbonates of Divalent Bases — Calcite Group

Calcite, $CaCO_3$.

Of all the minerals favored by collectors, calcite is of the widest and most common occurrence. In compensation for lack of rarity, calcite may be found in a bewildering variety of crystalline forms. Calcite provides a similar compensation to the fluorescent collector, occurring in a large variety of fluorescent colors, some of startling beauty.

The fluorescent calcites from Franklin and the Sterling mine at Ogdensburg, New Jersey, are probably the best known. In ordinary light, they resemble any other calcite from the extensive Franklin marble formation, being white and slightly translucent, and cleavable into rhombs. Away from the ore body, such calcite rarely fluoresces, but when this calcite has formed in close proximity to the ore at either location, a seemingly magical transformation takes place. Under short wave ultraviolet, such specimens glow a fiery orange-red, so much like a hot coal that beginning collectors sometimes will not pick up the specimen.

Calcite from Franklin is usually brighter under short wave than that from Ogdensburg. Under long wave, the response from both locations is quite variable, usually weakening to a pink or a dull red. Fluorescent calcite is most commonly associated with green fluorescing willemite, providing what is probably the most widely known and attractive fluorescent mineral combination. It may also be found with any of the other outstanding fluorescent minerals from Franklin, including hardystonite, esperite, wollastonite, or at Ogdensburg with sphalerite.

The activators at work in this red fluorescing calcite are generally well understood and owe their origin to the ore body. The red glow is caused by manganese atoms in substitution for certain calcium atoms. To this extent, calcite depends on the same activator as does willemite. However, the fluorescence activation of calcite is more complicated than this may indicate. Neither calcite nor the manganese activator can take up ultraviolet. As a result, for the red fluorescence to occur, a coactivator or sensitizer must also be present. Tests on a number of red fluorescing calcites indicate that lead atoms are the sensitizer for short wave ultraviolet. These atoms capture the short wave ultraviolet and transfer the energy to manganese atoms by a process of resonant transfer. This same combination of activators, manganese and lead, is the seat of fluorescence in a number of diverse minerals including apatite, wollastonite, and halite. In calcite, the brightest red fluorescence occurs at a manganese concentration of about 3% and a lead concentration of about 1%, though a fraction of this amount will be sufficient to produce fluorescence.

Lead is not the sensitizer for long wave fluorescence, however, for which the coactivator has not been firmly identified, though the visible light produced is again generated by manganese atoms.

Orange-red fluorescing calcites occur elsewhere, and when the occurrence is associated with nearby heavy metal ores, it is likely that the same activators are at work. Such red fluorescing calcite may be found with green fluorescing willemite at Moctezuma Canyon, Arizona; at Holcomb Valley, California; at the Kofa mine near Yuma, Arizona; and at numerous other locations. The calcite-willemite specimens appear indistinguishable from Franklin calcite and willemite specimens under ultraviolet light. At the Kofa, red fluorescing calcite may also be found with violet-blue fluorescing fluorite. This combination of fluorescent calcite and fluorite may be found at the Rat Tail claim in Arizona, sometimes with pink-orange fluorescent wickenburgite. Fluorescent calcite and fluorite masses are found at the American Tunnel in Silverton, Colorado. Here, calcite is also found as pink fluorescing crystals scattered on large, frosty crystals of quartz to form attractive specimens. Red fluorescent calcite is found, strangely, within magnetite at French Creek, Pennsylvania. It is also found as attractive red fluorescing crystal clusters under short wave at the Kelly mine at Magdalena, New Mexico, sometimes with sky blue fluorescing hydrozincite. All of these sites appear to be associated with heavy metal ores, and where metals have been hydrothermally deposited in calcite, red or red-orange fluorescing calcite is to be expected.

Bright orange-red fluorescent calcite cleavages are also found in the Tres Hermanos Mountains of New Mexico. Diamond shaped phantoms of bright red fluorescent calcite within clear calcite were once obtained at Ludlow, California. Red fluorescing calcite sandwiched between green fluorescing chalcedony is found at Hot Springs and is reported from Salmon, both in South Dakota. Red fluorescing calcite is also found in cracks, voids, and fissures in trap rock formations. For example, white crystals fluorescing pink-red are found in the diabase at Glenn Mills in Pennsylvania, and at Snake Hill in Secaucus and in basalt at Paterson and Bound Brook, New Jersey. While such trap rocks are not ordinarily thought of as sources of heavy metals, small amounts are disseminated in the trap rock, and may provide both manganese and lead as activators. As may be seen, red fluorescing calcite occurs in so many places in the United States that a complete listing is impossible.

Once we leave red fluorescence and consider other fluorescent responses of calcite, a bewildering variety is evident. At the same time, very little is known with confidence about the activators at work in these other fluorescent calcites. For example, the basalts at Bound Brook, New Jersey, also yield clear, attractive, dogtooth crystals fluorescing blue-white, and plates of large, clear, rhombic crystals of calcite fluorescing yellow-white, both with strong phosphorescence. These occasionally show a red phosphorescent flash when the ultraviolet light is quickly removed. At a limestone quarry near York, Pennsylvania, calcite crystals up to 4 inches in length fluoresce and phosphoresce white under both long and short wave, while the roots of the crystals fluoresce in wavy bands of red and vermillion under short wave.

Clear, attractive, calcite crystal groups are found at Rosiclare, Illinois, which fluoresce and phosphoresce shades of white under both long and short wave. Yellowish crystals of various forms, fluorescing weak yellow-orange, are found with galena in the tristate district.

Very interesting and attractive calcite crystals are found at Maybee, Michigan, bearing a thick, solid, porcelain-like white coat. These are disposed over yellow sulfur crystals and blue celestite crystals. Under either short or long wave, the calcite fluoresces a brilliant, solid, snow white, sufficient to make the sulfur and celestite visible and colorful in the illumination. On removal of the light, the phosphorescence is an intense green-white.

Small crystals of calcite, tan in color, found at Birdsboro, Pennsylvania, fluoresce yellow-tan with some phosphorescence. Yellow crystals of calcite are found near Barstow, California, fluorescing pale greenish yellow. Sun-parched tan calcite masses can be found throughout the deserts of California and Arizona fluorescing yellow, yellow-orange, or orange on the exposed surface. These thick coatings are often found with green fluorescing hyalite. At tungsten mines at Atolia and in the Shadow Mountains of California, such fluorescing calcite and hyalite may be found with blue fluorescing scheelite to form very attractive two- and three-color fluorescent specimens.

A variety of calcite known as "strontiocalcite" is found as colorless or tan crystals in vugs and cavities in the Faylor quarry at Winfield, Pennsylvania. This location is famous for fluorescing strontianite. The name strontiocalcite presumably means that the calcite contains substantial amounts of strontium. The fluorescence of this calcite is a bright, intense yellow, best under long wave, with a bright, long-lived, yellow-green phosphorescence. An apparently similar strontian calcite, with a similar but less intense fluorescence, has been found at Chittenango Falls, New York.

Cleavages of "Sky Blue" limestone (blue in daylight) come from the famous Crestmore quarry in Riverside, California. These sometimes fluoresce an interesting and attractive pink-orange under short wave, quite different in color from the red or red-orange response of lead-manganese activated calcite as typified by Franklin material. Some limestone from the same quarry is composed of crystals of calcite in granular, foliated masses. Many of the layers fluoresce a strong sky blue under short wave.

A very interesting fluorescent calcite has been found at Williamsport, Pennsylvania. This consists of a group of glassy crystals sprinkled with microscopic clusters of pyrite. Some of the calcite crystals fluoresce red while others fluoresce violet-blue, providing a very attractive color combination. Calcite is also found at Mimbres, New Mexico, fluorescing and phosphorescing deep sky blue under short wave. This is identical in response to the Terlingua calcite (discussed below) except that there is no pink response under long wave. A similar blue response is reported in calcite from Marple Falls, Texas.

We now consider what are undoubtedly some of the most interesting of mineral fluorescences – those in which the color differs depending upon whether short wave or long wave is used, as well as those in which the fluorescent and the phosphorescent colors differ. While such effects are found in many different mineral species the phenomena finds its greatest expression in calcite.

Undoubtedly the most famous such materials are the calcites found at the mercury mines at Terlingua, Brewster County, Texas. This locality also produces fluorescent calomel and terlinguite, which is a possible fluorescent. Under short wave, the calcite is a brilliant and intense sky blue or violet-blue. The blue fluorescence can be seen to build up slowly and methodically, starting with a brief pink or orange flash, until it reaches its full intensity. After the light is removed, there remains a bright blue phosphorescence of long duration. Under long wave, an astonishingly different response occurs. This is a bright pink fluorescence, sometimes with a slight violet tint. When the light is removed, the pink is immediately replaced by a blue phosphorescence of the same color but lesser intensity than that produced by short wave. Experiments with multi-wavelength sources indicate that wavelengths shorter than 3300A produce the blue fluorescence, while wavelengths longer than this produce the pink response.

This calcite is commonly mixed with a second type of calcite which is tan in ordinary light. This second calcite fluoresces and phosphoresces bright yellow under both short and long wave. A calcite specimen from Terlingua may thus produce blue and yellow fluorescence and phosphorescence under short wave, and under long wave pink and yellow fluorescence followed by blue and yellow phosphorescence. As a bonus, some specimens are found with hyalite which fluoresces green under short wave.

It is probably not widely recognized, but calcites in most respects identical in fluorescent response to Terlingua calcite are found at other places such as Sabina, Ohio; Shelby, Indiana; San Saba in Texas; and Apache, Oklahoma. An apparently similar fluorescent calcite is also reported from Hurley, New Mexico and Miles, Texas. The San Saba material provides a brief yellow flash before the onset of steady pink under long wave. Further, the blue is distinctly more pink than Terlingua material. The response is in other ways identical.

A quarry in Monroe County, Pennsylvania, provides a calcite resembling the Terlingua material. This fluoresces an intense sky blue under short wave, the color being indistinguishable from that of the Terlingua material. As in the Terlingua calcite, the fluorescence buildup to full intensity is slow and the phosphorescence is long lasting. However, under long wave, the response is yellow rather than the pink of the Terlingua material. A generally similar calcite is found near Gouverneur, New York. The short wave fluorescence is pink-orange, but the phosphorescence is the exact blue color of the previously mentioned calcites. Apparently, in this case, the orange fluorescence masks a concurrent blue response which is then seen when the ultraviolet light is turned off.

In the case of many of the blue fluorescent calcites mentioned, the portion of the material which fluoresces has a distinctly pink color in daylight which will slowly change to green after sustained exposure to daylight. Sometimes specimens are found which are greenish on the side long exposed to daylight and pink on the sheltered underside. No difference in the fluorescence is evident, however. Such calcites present a challenge to scientific explanation, combining as they do different responses under short and long wave ultraviolet, and associated as they often are with color centers. Little research has been done on this matter. Certain recent investigation suggests, however, that the blue fluorescence may be due to europium or neodymium found to be present.

A calcite obtained from Hurricane, Utah, provides a particularly beautiful two-color fluorescent response. This material has a banded structure composed of small, clear, calcite crystals layered between dense, white calcite. Under short wave, alternate bands fluoresce a warm orange-red, while the remaining material responds in an indisinct and cloudy purple. Under long wave, the formerly red bands now fluoresce bright yellow, while the other portions continue faint purple. When the light is removed, the once purple areas now phosphoresce briefly in blue-white. This calcite is sometimes accompanied by blue fluorescent hydrozincite coatings which add to its attractiveness.

Calcite in very thin, clear, colorless, bladed crystals is found in the Badlands of South Dakota. The long wave response is white, but under short wave, the response is green, suggestive of a uranium activator in very small concentrations. A very similar calcite is found in the Buckskin Mountains, Yuma County, Arizona.

Back in the Franklin mining district in New Jersey, a calcite from the ore body at Ogdensburg has been found which fluoresces a weak dark red under short wave. Under long wave, some portions continue to fluoresce red, but other segments now fluoresce a light yellow. A far more attractive, but not well-known, two-color fluorescent has been found on Buckwheat dump at Franklin. Under short wave, this fluoresces bright pink-red similar to much of the Franklin calcite. Under long wave, however, the response is a vivid and unusual violet-blue. At Ogdensburg, calcite is also found as small, glass-clear, pyramidal crystals on limestone, locally called "water drop" calcite. Some of these fluoresce pale yellow under either short or long wave light, but on removal of the light there is a distinct red-orange flash followed by a slow yellow phosphorescence, similar to that occasionally seen in calcite from the quarry at Bound Brook, New Jersey.

Magnesite, $MgCO_3$.

Magnesite is one of the several fluorescent minerals which may be found in the serpentine quarries on the Maryland-Pennsylvania border. These quarries were mined for chromite and were the primary domestic source of this strategic metal during the First World War. The magnesite is an alteration product of the host serpentine rock and occurs as white or yellowish coatings or porcelain-like masses.

The fluorescence is usually a bright blue-white, best under long wave. In some specimens, there is a phosphorescence of the same color. In one specimen, the generally white phosphorescence was preceded by a momentary red-orange flash, much as seen in some calcite. Magnesite with a similar fluorescence is reported from Hoboken, New Jersey, and from Kern County, California.

Smithsonite, $ZnCO_3$.

Smithsonite is occasionally fluorescent. For the most part, its fluorescence is undistinguished, with the material from Franklin and Ogdensburg, New Jersey, being the most attractive. At Franklin, smithsonite masses fluoresce pale yellow under long wave. What is believed to be smithsonite is found frequently on the dumps as white coatings on calcite, while at the Sterling mine at Ogdensburg, a white vein material is found, also believed to be smithsonite. This fluoresces a lovely orange-yellow of egg yolk color, best under long wave. Smithsonite after dolomite from Rush, Arkansas, fluoresces orange under long wave, while material from Granby, Missouri, may fluoresce yellow or pink, also under long wave ultraviolet.

Smithsonite in the form of small crystal druses is found at Rush and at the Jackson Tunnel in Luna County in New Mexico. These samples fluoresce dull white or gray-white under long wave. A similar fluorescent response is reported in smithsonite from Joplin, Missouri; the Cerro Gordo mine in California; and Marion County, Georgia.

Carbonates of Divalent Bases — Aragonite Group

Aragonite and Nicholsonite, $CaCO_3$.

Aragonite is a dimorph of calcite, an alternative form of calcium carbonate. The change in molecular design produces an apparent change in the fluorescent potentiality of calcium carbonate, since the variety of fluorescent responses seen in aragonite is distinctly more limited than that seen in calcite.

Aragonite was found in the zinc workings at Friedensville, Pennsylvania, as white, radiating blades completely filling fissures in the host rock. The fluorescence is white coloring to yellow in portions of the blades, best under long wave. This aragonite provides a long-lived phosphorescence of a green-white color. Similar, long, radiating, glassy blades of aragonite are taken from the Sterling mine at Ogdensburg, New Jersey, with much the same fluorescence. Clusters of free-standing pointed crystals of aragonite are found at San Luis Obispo, California, with fluorescent response similar to the Friedensville material. A similar fluorescent response is reported from locations too numerous to list, but including Magnet Cove, Arkansas; Elizabethtown, Illinois; Bisbee and Kingman, Arizona; and Crestmore, California. We can consider this white or yellowish white fluorescence with strong phosphorescence the typical response of aragonite.

Aragonite is found as snow white clusters of free-standing crystals at Gold Hill, Utah. These fluoresce an intense green of the color of a uranium activated mineral, most vivid under short wave but certainly quite bright under long wave. It is probably safe to assume that the activator is indeed uranium. A similar fluorescence is reported in aragonite from Organ and Tres Hermanos, both in New Mexico.

Nicholsonite is a zinc bearing aragonite. Material from the zinc mine in Friedensville, Pennsylvania, was found as powdery white masses. Under short wave, the fluorescence is strong white with white phosphorescence. However, under long wave, the fluorescence is a beautiful, strong orange-yellow with a similar phosphorescence. The contrasting performance under the two lights is quite impressive.

Additional information on aragonite will be found in Part B of this chapter.

Cerussite, $PbCO_3$.

Cerussite is the most frequently encountered of the lead minerals which occasionally fluoresce. The typical fluorescent color is yellow or yellow-orange, strongest under long wave ultraviolet. Fluorescent crystals of cerussite are found on the dumps at Phoenixville and at Audubon, Pennsylvania, in a stubby bladed habit, fluorescing yellow or yellow-orange under long wave and weaker under short wave. The mine at Audubon was owned by the father of J. J. Audubon, the renowned painter of birds, and the younger Audubon spent much of his youth there. This mine operated in a vein, predominantly of lead minerals. Close by, other shafts exploited a copper vein. Together with the nearby Phoenixville mines, these supplied a variety of supergene minerals, perhaps among the most varied in the eastern United States.

Cerussite crystals in a jackstraw habit are found at Kellogg, Idaho; these sometimes fluoresce yellow-white or sometimes bright yellow. The jackstraw cerussites from Tiger, Arizona, also sometimes fluoresce yellow-white, and the magnificent jackstraw crystal groups taken from the Flux mine in Santa Cruz County, Arizona, will infrequently fluoresce weak yellow-white under long wave. Cerussite from the Sterling mine at Ogdensburg, New Jersey, fluoresces yellow under long wave.

The fluorescences of the lead minerals cerussite, anglesite, leadhillite, matlockite, and phosgenite generally resemble one another. When these fluoresce at all, the response is typically yellow, sometimes more orange in shade, sometimes more white, and strongest under long wave ultraviolet. This suggests that some common activator is at work, possibly the lead itself. The similarity of the fluorescent response makes these minerals difficult to distinguish from one another on the basis of fluorescence.

Strontianite, $SrCO_3$.

Fluorescent strontianite is found plentifully at the Faylor quarry in Winfield, Pennsylvania, as white tufts of small crystals often mixed with larger, clear crystals of nonfluorescent calcite. The strontianite fluoresces bright white with a blue cast, best under long wave. The phosphorescence varies, but sometimes is an intense and enduring greenish white. Similar fluorescent strontianite is reported in a number of other quarries in the area and in Schoharie County, New York.

Strontianite is also found at the Minerva mine at Cave-in-Rock in Hardin County, Illinois, as large, tapered, jackstraw crystals. These fluoresce yellow-white, best under long wave with little sign of phosphorescence. Strontianite crystals fluorescing peach or pink under long wave have been found in the East Coleman District, Furnace Creek, Death Valley.

Additional information on strontianite will be found in Part B of this chapter.

Witherite, $BaCO_3$.

Found at Rosiclare, Hardin County, Illinois, sometimes with fluorite for which this location is famous, witherite is sometimes recognizable by its great weight due to the barium in its composition. The mineral is found as white bipyramidal crystals or ball-like masses composed of layers of flat hexagonal crystals. The best fluorescence is an intense white with a blue cast especially strong under long wave ultraviolet. In other specimens, the color is a decided yellow-white. Brightness varies greatly among specimens. Witherite usually has an intense and enduring phosphorescence, yellowish in color.

Witherite, said to be from quarries in Woodville, Ohio, fluoresces like the Illinois material. Witherite is also found as white or clear tabular crystals at the Pigeon Toe mine in Montgomery County, Arkansas; these fluoresce yellow-white, best under long wave ultraviolet.

Barytocalcite and Alstonite, $BaCa(CO_3)_2$.

Several barium calcium carbonates of apparently identical chemical composition but different molecular structure exist, including barytocalcite and alstonite. All are elusive minerals and all are reported as fluorescent. Barytocalcite is reported from Lewisburg, Pennsylvania, fluorescing weak yellow with a greenish afterglow. Alstonite is known to be a fluorescent and has been reported from the Minerva mine at Cave-in-Rock fluorescing yellowish white under long wave. An interesting member of the barium calcium carbonate group, as yet unnamed, is reported from Cave-in-Rock as a coating on barite and fluorescing a pale to bright orange under long wave.

Carbonates of Mono-and Divalent Bases

Shortite, $Na_2Ca_2(CO_3)_3$.

Shortite occurs as colorless, clear crystals in a brown matrix material from the FMC mine, Green River, Wyoming. The crystals fluoresce bright yellow and the matrix material brown under long wave. The response is similar but weaker under short wave.

Pirssonite, $Na_2Ca(CO_3)_2 \cdot 2H_2O$.

Small crystals of pirssonite are sometimes available from Searles Lake, California. These fluoresce under both long and short wave. The color is a not very bright blue-white to blue-gray, with a strong but short phosphorescence.

Gaylussite, $Na_2Ca(CO_3)_2 \cdot 5H_2O$.

Gaylussite is reported from the state of Washington as fluorescing pink of moderate intensity under long wave; from Searles Lake in California as fluorescing weak orange under long wave; and from Ragtown Lake, Nevada, as fluorescing pale cream-white under short wave. Specimens of gaylussite from Searles which have been examined appear to owe their weak orange or pink fluorescence to lake mud. The source of the fluorescence of the mud is not known.

Additional information on gaylussite will be found in Part B of this chapter.

Carbonates of Other Types

Hydrozincite, $Zn_5(CO_3)_2(OH)_6$.

The fluorescence of hydrozincite is usually an intense sky blue under short wave, in some specimens with a touch of violet, in other specimens with a slightly whiter shade. Usually, this mineral is nonfluorescent under long wave, but occasionally hydrozincite specimens are found with a cream or dull violet fluorescence under long wave. It is not established whether this is due to the hydrozincite itself or to the presence of other accompanying carbonates. The sky blue fluorescent color of hydrozincite is virtually identical to that of some scheelite, but hydrozincite may be distinguished by low hardness, powdery appearance, and associations.

Hydrozincite is found at Franklin and Ogdensburg, New Jersey, fluorescing bright, deep, sky blue, and coating red fluorescing calcite. Attractive, large masses of this blue fluorescent material were found on mine dumps in Audubon and Phoenixville, Pennsylvania. It is found also in large blocks and masses at Goodsprings, Nevada. Hydrozincite can be found on the mine dumps at numerous places in the metalliferous belt of the West as a weathering product of sphalerite.

While hydrozincite is usually a soft, powdery material, dense compact masses of hydrozincite have been found deep in the mine at Franklin, New Jersey, fluorescing an unusual dark purple-blue.

Additional information on hydrozincite will be found in Part B of this chapter.

Dypingite, $Mg_5(CO_3)_4(OH)_2 \cdot 5H_2O$.

A hydrated magnesium carbonate from the Sterling mine in Ogdensburg, New Jersey, the fluorescent response of dypingite is reported to be gray-blue under short wave and light blue under long wave.

Monohydrocalcite, $CaCO_3 \cdot H_2O$.

The hydrated calcium carbonate, monohydrocalcite, is found in the Sterling mine with the appearance of dried, runny, white glue. Under the short wave light, the response is a vivid green, similar in color to the fluorescent response of a uranium mineral such as andersonite. The long wave response is similar but weaker.

This mineral is probably of wide occurrence and may be found wherever calcium rich waters trickle over exposed surfaces or on surfaces upon which sprays deposit calcium minerals. Fluorescence may help to determine its presence.

Benstonite, $(Ba, Sr)_6(Ca, Mn)_6Mg(CO_3)_{13}$.

A carbonate of magnesium, calcium, barium, and strontium, benstonite from the Barold mine, Magnet Cove, Arkansas, is found as dense white masses with glistening, randomly oriented facets. Under either short or long wave, much of the material fluoresces pink-red, with a brief red phosphorescence. The general effect is that of a fluorescent calcite, and the same manganese activator may be at work here. Benstonite is found at a fluorite mine at Elizabethtown, Illinois, fluorescing pale white and at the Minerva mine at Cave-in-Rock, Illinois, with a weak yellow fluorescence.

Northupite, $Na_2Mg(CO_3)_2Cl$.

Northupite, found as sharp bipyramidal crystals, is an occasional fluorescent from Searles Lake at Trona, California. The fluorescence is bright cream-white under long wave ultraviolet, with strong phosphorescence.

Tychite, $Na_6Mg_2(CO_3)_4(SO_4)$.

Tychite, also found at Searles Lake, is related to northupite and is similar in crystal form, though the crystals are smaller. Fluorescence is cream-white, best under long wave.

Whewellite, $CaC_2O_4 \cdot H_2O$.

Whewellite from Meade County in South Dakota fluoresces a strong blue-white under both short and long wave ultraviolet.

BORATES

Hydrous Borates of Monovalent Bases

Tincalconite, $Na_2B_4O_7 \cdot 5H_2O$.

Tincalconite from Boron, California, is snow white in color and occasionally will fluoresce white under either short or long wavelength.

Borax, $Na_2B_4O_7 \cdot 10H_2O$.

Borax is unstable and, when taken from its natural environment, may alter to white coatings of tincalconite by loss of water. Fresh, unaltered fragments of borax from Searles Lake may fluoresce weakly under long wave, the color being yellow-white.

Hydrous Borates of Mono- and Divalent Bases

Ulexite, $NaCaB_5O_9 \cdot 8H_2O$.

Ulexite is found plentifully in the open pit at Boron, California, as tight bundles of clear, columnar, glassy blades. When the ends of specimens of this kind are polished to optical flatness, they can act as "fiber optics." If one flat polished end is placed in contact with a picture, the picture appears as if on the top surface of the bundle. Because of this effect, such material is called "TV stone" since the top then looks like a miniature TV screen. Ulexite is more widely known for this property than for fluorescence. The Boron material fluoresces only occasionally in a weak yellow-white under either long or short wave, with some phosphorescence.

Probertite, $NaCaB_5O_9 \cdot 5H_2O$.

Probertite is a major borate ore at the Billie mine in Death Valley. Here, it is found as tight bundles of long bladed tan crystals. Sometimes areas of butter yellow or lemon yellow fluorescence under short or long wave are found in probertite. As with other borates, there is always the possibility that the material seen is an alteration, except that fluorescence was quite evident on freshly exposed surfaces.

Hydrous Borates of Divalent Bases

Tunellite, $SrB_6O_{10} \cdot 4H_2O$.

A very interesting strontium borate found at Boron, California, tunellite is sometimes distinguishable from the other borates with which it may be found through the pearly appearance which it may present or through its distinctive diamond shaped crystals. The fluorescence is butter yellow under long or short wave. Since the fluorescence is uneven, it may be due to an alteration.

Colemanite, $Ca_2B_6O_{11} \cdot 5H_2O$.

Colemanite is of wide occurrence in the boron mineral ores of California as clear, colorless, arrowhead shaped crystals or as white or tan blocky crystals, rhombs, cleavages, or masses. In specimens from Death Valley or Boron, the colorless crystals seem to have the weaker fluorescence, usually in a pale blue-white or pale yellow color. The more deeply colored and less transparent material provides the stronger fluorescence. The fluorescent color may be pale white, bright butter yellow, golden yellow, or lemon yellow, and is usually strongest under long wave. Colemanite often shows an exceptional phosphorescence in terms of, intensity and duration, the phosphorescent color ranging from light yellow-white to green-white.

Most attractive specimens of colemanite have been found in the pit at Boron. These are arrowhead shaped crystals of colemanite which fluoresces white, disposed artistically over yellow calcite which fluoresces yellow. In the Thompson mine or the Boraxo No. 1 pit in Death Valley, blocky colemanite crystals of extraordinary size were found at the pit's bottom. These may fluoresce yellow-white, blue-white, or weak orange under long wave.

Gowerite, $CaB_6O_{10} \cdot 5H_2O$.

The rare mineral gowerite is found as very fine, silky, acicular crystals at the Mott Prospect, Death Valley, California. The fluorescence is a weak white or blue-white, almost imperceptable in brightness.

Meyerhofferite, $Ca_2B_6O_{11} \cdot 7H_2O$.

Meyerhofferite is found at Death Valley, California, as clear, colorless, tabular crystals, sometimes with an alteration to a white material at crystal edges. The meyerhofferite fluoresces yellow-white under long wave, while the apparently altered areas fluoresce butter yellow under both long and short wave. This alteration may be to colemanite. Meyerhofferite from the Monte Blanco mine in Death Valley is occasionally seen with an intense lemon yellow fluorescence, best under short wave but still bright under long wave.

Like most of the borates, the fluorescence of meyerhofferite is quite variable, probably due to alteration associated with different states of hydration or to other chemical changes.

Priceite, $Ca_4B_{10}O_{19} \cdot 7H_2O$.

Priceite from the Thompson or Boraxo Pit in Death Valley, California, fluoresces white, best under long wave.

Hydroboracite, $CaMgB_6O_8(OH)_6 \cdot 3H_2O$.

Hydroboracite from the Thompson shaft in Death Valley, California, fluoresces variably. Some specimens will fluoresce white, and others will fluoresce blue-white or yellow-white. The response is best under long wave. As with other borate minerals from Death Valley, it is difficult to be certain that the mineral identified is really the fluorescent, since these minerals are subject to extensive alteration due to changes of water content and other causes.

Anhydrous Borates of Divalent Bases

Szaibelyite, $MgBO_2(OH)$.

A rare and not particularly attractive mineral, szaibelyite is found as a soft white matted coating on green serpentine at Stinson Beach, California. The fluorescent response is a weak yellow-white under long wave. Tan sugary material in host rock from the Crestmore quarry in Riverside, California, is said to be szaibelyite. This fluoresces tan under both short and long wave ultraviolet.

Szaibelyite forms a continuous series with the well-known Franklin mineral sussexite. While sussexite contains manganese which szaibelyite ideally does not, and while manganese is a widespread activator, sussexite does not fluoresce.

Fluoborite, $Mg_3(BO_3)(F,H)_3$.

The rare mineral fluoborite is found in the Franklin marble near Hamburg, New Jersey, as small white crystals in nonfluorescent white calcite. When the calcite is broken open, the crystals of the fluoborite are usually exposed undamaged and look like small protruding grains of rice. The fluorescence is a moderately strong yellow under short wave. Norbergite is also found in these same calcites and fluoresces very similarly; it is somewhat difficult to distinguish from fluoborite on the basis of fluorescent response. The rice-grain appearance of the fluoborite provides the best quick basis by which to distinguish between these minerals. Fluoborite has long been known from the Sterling mine at Ogdensburg, New Jersey. This material has a fibrous appearance and appears to be nonfluorescent. However, recently, fluorescing fluoborite has been reported from the Sterling mine, the reported response being yellow-white under long wave.

SULFATES

Anhydrous Sulfates of Monovalent Bases

Aphthitalite (Glaserite), $(K, NA)_3Na(SO_4)_2$.

From Searles Lake in California, aphthitalite fluoresces butter yellow under long wave, weaker under short wave ultraviolet.

Thenardite, Na_2SO_4.

Large gray-white crystals of thenardite are found near Sodaville in Mineral County, Nevada. These crystals fluoresce bright white under either short or long wave ultraviolet. Removal of the light reveals a strong and enduring greenish white phosphorescence. A generally similar response is reported from material from Nogales, New Mexico.

Anhydrous Sulfates of Divalent Bases — Barite Group

Celestite, $SrSO_4$.

Hexagonal, flat topped crystals of celestite with white jackets and blue cores to about an inch or two in length are found at Lime City quarry and at the Pugh quarry in Weston, Ohio. These fluoresce bright white under both short and long wave, but best under long wave. The response seems strongest on certain preferred faces of the crystal. Phosphorescence is sometimes strong, usually green-white in color. It has been suggested that the white jacket is an alteration to strontianite. However, X-ray analysis performed on such material from the Lime City quarry showed only celestite with no trace of strontianite. Similar response is reported in celestite from Portage and Clay Center, Ohio. Celestite from Wymore in Nebraska fluoresces white under both long and short wave.

Fluorescence of cream-white color is also reported from some celestites from Walford, Cazenova, and Chittenango Falls, all in New

York State, but the reports differ on whether the short or long wave response is best.

Celestite is found at the Sterling mine in Ogdensburg, New Jersey, in thin veins or as small tan crystals in vugs. These fluoresce weak yellowish white or bluish white, best under long wave.

Anglesite, $PbSO_4$.

The old mine dumps of the Wheatley mine at Phoenixville, Pennsylvania, produce numerous secondary minerals, including fluorescent cerussite and anglesite. The fluorescence of this anglesite is a bright yellow or yellow-orange, best under long wave ultraviolet. This fluorescent response is identical to that of the cerussite, and it is impossible to distinguish between these minerals from this location on the basis of fluorescence.

This yellow or yellow-orange response is typical of anglesite and is found in anglesite from the Bunker Hill mine at Kellogg, Idaho. However, in some of these Kellogg anglesite crystals, an unusual and attractive pink or lavender fluorescence is shown, brightest under long wave. In some specimens, some crystals will fluoresce the usual yellow, while others will fluoresce bright pink.

Barite, $BaSO_4$.

Barite is a frequent fluorescent, most commonly displaying a not very saturated or intense light yellow color. Barite is found as beautiful and showy crystals in several places in South Dakota, particularly in the area of the Cheyenne and Moreau rivers. The barite crystals are tan to brown, glassy and transparent. Under short wave ultraviolet, these barites fluoresce butter yellow. Often there is a phosphorescence, usually a strong green-white. Accompanying calcite also fluoresces yellow, but more orange than the barite, and the combination of barite and calcite makes an attractive fluorescent combination.

Barite is found as white to tan flat crystals standing on fluorite from the Buckskin Mountains, Yuma County, Arizona. This barite fluoresces orange-tan under long wave ultraviolet, while the fluorite responds with its usual blue, forming another attractive two-color fluorescent combination. Barite is also found at Bingham, New Mexico, as thick, white, blady crystals with fluorite. In these specimens, the barite fluoresces deep pink or weak red while the fluorite fluoresces blue, under long wave. Weak red fluorescing barite with fluorescing fluorite is also found at Elmwood, Tennessee.

Barite is a well-known Franklin fluorescent. It is found as blebs or veins in the host calcite. Typically, the barite fluoresces strong yellow or yellow-white under short or long wave, and typically also, the calcite fluoresces its familiar red-orange. Because of its color and intensity, this combination is much sought after by collectors. Interestingly, while the barite usually fluoresces under both long and short wave, in some few Franklin specimens there is no appreciable response under long wave. In other specimens, the barite will fluoresce yellow under long wave and blue-white or blue-gray under short wave. Rarely also, Franklin barite will fluoresce violet, strongest under long wave. The fluorescent response of barite from Franklin is thus quite variable.

One of the most sought after fluorescent Franklin specimens is the combination of wollastonite with calcite in their fluorescent oranges and reds. Sometimes such specimens include small spots or blebs of blue-gray fluorescing barite, which contributes an interesting color accent.

Barite is found at Cave-in-Rock, Hardin County, Illinois, as transparent or white arrow-like crystals. These fluoresce butter yellow under short or long wave, sometimes brightly, and with some phosphorescence. Similar fluorescing barite is found at Pugh quarry in Weston, Ohio, and at Palos Verdes, Los Angeles County, California; it is also reported from Magnet Cove, Arkansas.

In the past, material was obtained from Hot Springs, North Carolina, as bright orange fluorescing bands and veins under long wave, in a nonfluorescing gray host rock. This was long said to be barite, but recently it has been said in fact to be sphalerite.

Additional information on barite will be found in Part B of this chapter.

Hydrous Sulfates of Divalent Bases

Selenite (Gypsum), $CaSO_4 \cdot 2H_2O$.

Selenite crystals from a number of locations show a white fluorescence under long wave, which is weaker under short wave. This white is seen as having either a bluish or yellowish tint depending on conditions of observation. This fluorescence is generally followed by a bright and enduring green-white phosphorescence once the light is removed. Selenite from clay banks along the Chesapeake Bay, Maryland, and from Cody, Wyoming, shows this response.

Selenite crystals are often found in the form of rhombs. If we imagine two lines drawn diagonally from corner to opposite corner, the crystal will be divided into four triangular regions by these two diagonals. If the crystal is transparent, within these triangles a remarkable fluorescent phenomenon is often seen. The triangle within the crystals based along the long edge of the crystal will fluoresce as will the triangle directly opposite, but the other two triangles of the crystal will be nonfluorescent! This fluorescence in the shape of an "hourglass" can be seen very well in selenite from Cody.

Fluorescent selenite is reported from numerous other localities, including Franklin, New Jersey; Gypsum, Ohio; and Lovington, New Mexico.

Hydrous Sulfates of Di- and Trivalent Bases

Ettringite, $Ca_6Al_2(SO_4)_3(OH)_{12} \cdot 26H_2O$.

Ettringite is a very rare mineral at Franklin, New Jersey. It is reported to fluoresce yellow-white under both short and long wave ultraviolet.

Sulfates with Other Acids

Sulfohalite, $Na_6(SO_4)_2FCl$.

Sulfohalite is a scarce mineral found as bipyramidal crystals up to one inch long at Searles Lake, California, fluorescing bright white or cream-white under long wave ultraviolet. This material is often phosphorescent.

Burkeite, $Na_6(CO_3)(SO_4)_2$.

Burkeite is found at Searles Lake in California. This material may fluoresce yellow-white under short or long wave ultraviolet, with some phosphorescence.

Hanksite, $Na_{22}K(SO_4)_9(CO_3)_2Cl$.

Hanksite is found at Searles Lake in fat, colorless or slightly gray-green crystals sometimes of astonishing size. Fluorescence is a weak to moderate pale yellow, sometimes zoned, and with a slight phosphorescence. The response is slightly better under long wave.

Leadhillite, $Pb_4(SO_4)(CO_3)_2(OH)_2$.

The rare lead mineral leadhillite is found in vugs and cavities, often associated with galena, at the Beer Cellar mine, Granby, Missouri. The fluorescence is an intense yellow, stronger under long wave than short. Leadhillite is a well-known rare mineral from Tiger in Arizona, but these specimens only occasionally fluoresce a weak gray under either short or long wave ultraviolet.

Creedite, $Ca_3Al_2(SO_4)(F,OH)_{10} \cdot 2H_2O$.

Creedite from a fluorite mine at Wagon Wheel Gap, Colorado, is reported to fluoresce cream-white under long or short wave, or pale blue under short wave according to another report. These could well be identical fluorescent responses seen under different lighting conditions or by different eyes.

TUNGSTATES AND MOLYBDATES

Scheelite, $CaWO_4$.

Scheelite is an undistinguished material in daylight appearance, often easily mistaken for milky quartz or some other common rock or mineral. However, in the dark under the short wave light, scheelite is a star performer, often fluorescing in an almost unmistakable, intense sky blue. The blue response, which is found in the purest grades of sheelite, is only seen under short wave. When impurities are present, a white, yellow, orange-yellow, brown, and even faintly pink response may be seen. These responses are strongest under short wave, but occasionally show under long wave.

Molybdenum is a common substitute for some of the tungsten atoms in scheelite, forming a series toward powellite. As little as one-third of one percent of molybdenum will produce the white response, and over one percent will produce a yellow response. A theory explaining this change of fluorescent color due to the presence of molybdenum has been developed by Russian scientists. This theory is based on investigation of the physical properties and measured color change in synthetic scheelites with various amounts of molybdenum. At well under 1.6% of molybdenum, the atoms of molybdenum are scattered through the scheelite, replacing tungsten atoms randomly. At about 1.6% molybdenum, the molybdenum atoms are no longer randomly distributed but begin to cluster in certain crystallographic planes. This produces a local powellite environment in one plane, and the fluorescent response moves toward longer wavelength, that is, toward yellow. At about 15% molybdenum, the clustering becomes two dimensional, so that certain regions within the specimen become small scale pockets of powellite. This produces a true powellite yellow fluorescent color, which persists for all higher concentrations of molybdenum.

From the point of view of mineralogical nomenclature, the mixed mineral is called powellite only when molybdenum exceeds tungsten, that is, past the 50% point, but it is apparent that from the viewpoint of fluorescence, the crossover is reached at about 15% molybdenum.

When even a few percent of molybdenum is present, the commercial value of scheelite decreases, so the fluorescent color enables one to distinguish the more valuable ore from the less valuable ore. To aid in this type of assay, standard samples are available with known amounts of molybdenum which can be used for comparison under the ultraviolet light.

The mines at Atolia near Randsburg, California, have been one of the largest producers of very high grade scheelite. Scheelite has been taken out through innumerable small vertical shafts working the thin white veins of ore. Additional scheelite has been found through extensive near surface high grading in an area called the "spud patch." Such scheelite fluoresces an intense sky blue under short wave, attributable to its high purity. Gold is sometimes an accompaniment of scheelite and has been mined from sites at Randsburg. Both ores apparently owe their origin to hydrothermal activity accompanying a now extinct vulcano nearby. Elsewhere in the West, tailings which had been abandoned as worthless after the mining of gold proved to contain valuable scheelite, recognized only after the application of ultraviolet light revealed widespread fluorescence.

Not far from Atolia, in the Shadow Mountains, scheelite is found as blue fluorescing spots in bands running through the host rock. Here, specimens may be found containing orange fluorescing calcite and green fluorescing hyalite, providing very attractive three-color fluorescent specimens. Here also, scheelite which fluoresces a beautiful, intense, canary yellow under short wave, may be found.

Scheelite is reported at the Crestmore quarry at Riverside, California, reportedly fluorescing bright yellow-white under short wave. Scheelite as small yellow fluorescent spots under short or long wave is found rarely on the dumps at Franklin, New Jersey. Scheelite from the New Ortiz gold mine in Santa Fe County, New Mexico, and from Mill City, Nevada, fluoresces variously sky blue, white, or yellow, the latter two colors showing also under long wave.

The old mine at Trumbull, Connecticut, perhaps the earliest scheelite mine is the United States, produced scheelite which fluoresces a vivid sky blue in spots to one-half inch in size.

Fluorescent specimens of exceptional beauty are found in a small mine near Oracle, Arizona. These consist of a matrix of bright red fluorescing calcite through which run veins and wispy strands of blue fluorescing scheelite. Two-color fluorescing scheelites are also known. This seems to be a characteristic of some scheelites which are tan or brownish in ordinary light. These may fluoresce the commonly seen white or blue-white under short wave ultraviolet, but under long wave, certain of these specimens will fluoresce burnt orange or tan. Scheelite with this dual response is found in the Huachuca Mountains in southeastern Arizona.

Because of its industrial importance, many sources of scheelite are known, particularly throughout the western states, so that the above descriptions are merely representative of a far larger number of mines and prospects.

The blue fluorescence of scheelite has been extensively investigated, and no activator has been found to be responsible for this fluorescence. Indeed, synthetic scheelite of the highest purity has been prepared, and it will fluoresce blue. It is believed that scheelite may be self-activated with fluorescences due to charge transfer between tungsten and oxygen forms.

Powellite, $CaMoO_4$.

An analogue of scheelite, powellite is much rarer than scheelite with which it forms a substitution series. The fluorescence of powellite is a rich yellow under short wave. Powellite from Garlock, California, a Dana locality, fluoresces in this color. Specimens from Coaldale, Nevada, contain both powellite and hyalite. Such specimens fluoresce in a striking yellow and green combination. Powellite is reported from Franklin and Ogdensburg, New Jersey, as an alteration of molybdenite, fluorescing yellow. Since scheelite containing molybdenum also fluoresces yellow, it is likely that some misidentification exists among powellite and scheelite specimens.

Stolzite, $PbWO_4$.

Fluorescing stolzite is reported from scheelite deposits south of Tombstone, Arizona, and also from the vicinity of nearby Dragoon. It is also reported near Lucin, Utah; from the Wheatley mine in Phoenixville, Pennsylvania; and from Southampton, Massachusetts. In all of these cases, the fluorescence is reported to be green-white.

It is now considered doubtful that the Wheatley material is really stolzite. X-ray examination of specimens labeled as stolzite from this location has shown that these are really wulfenite. Such misidentification of wulfenite as stolzite is not surprising. The two minerals are part of a series in which tungsten and molybdenum can substitute for one another, and wulfenite and stolzite crystals have much the same appearance. While stolzite appears to be a confirmed fluorescent based on valid fluorescent material from two other continents, these overseas specimens fluoresce yellow, not green-white. These considerations leave the identification of fluorescent stolzite from these United States locations in some doubt. Some light may be cast on this by research done on synthetic materials. Synthetic stolzite fluoresces yellow, and the movement toward wulfenite by the substitution of some molybdenum for tungsten does not change this response. However, a movement toward scheelite by the substitution of some calcium for lead can lead to a green response. Thus, the reported green-white fluorescence may represent that of a calcium rich stolzite.

The series of minerals scheelite-stolzite-wulfenite-powellite merit further research in order to clarify their relationships in nature and the fluorescent responses to be expected at various points in the series.

Additional information on stolzite will be found in Part B of this chapter.

Wulfenite, $PbMoO_4$.

Wulfenite does not appear to be a fluorescent ordinarily, but some fluorescent specimens are found. Exploratory pits around Tiger, Arizona, yield wulfenites fluorescing orange, low to medium intensity, and some of the wulfenite from the Mammoth mine at Tiger fluoresces weak red-orange to burnt orange, best under long wave ultraviolet. Wulfenites from the Finch (Barking Spider) mine in Gila County, Arizona, fluoresce pale yellow under short or long wave, and sometimes green due to a thin hyalite coating.

Red-orange wulfenite crystals from the Red Cloud mine in Yuma County, Arizona, are outstandingly beautiful and world famous. These wulfenites occasionally fluoresce. The color is cherry red under long wavelength. The response is weak, but such red fluorescing crystals may be framed in yellow fluorescing aragonite and accompanied by bright blue fluorescing fluorite and green fluorescing willemite, providing an attractive combination of fluorescent minerals.

PHOSPHATES AND ARSENATES

Apatite Group

Apatite, $Ca_5(PO_4)_3F$.

Apatite can be found in the Franklin marble of northern New Jersey and southern New York State, and in similar marbles in northern New York State, as light blue, pale green, or colorless crystals. Such apatite, more exactly fluorapatite, sometimes fluoresces pale sky blue to white under short wave ultraviolet. Blue apatite crystals in calcite from Pitcairn, New York, provide an example. The apatite fluoresces white while the enclosing calcite fluoresces weak red. Apatite with a blue-gray fluorescence is found in quarries in the vicinity of Franklin, New Jersey. The activator at work in these white, pale sky blue, or blue-gray fluorescences has not been determined. Based on studies of synthetic apatite, it is known that small amounts of tin, antimony, or even copper can produce a bluish response in apatite and, with the addition of small amounts of manganese, a white response, but it is by no means clear that these are at work in the natural material. Synthetic apatites with these activators are used as phosphors for fluorescent lights.

Apatite is also found close to or within the ore bodies at Franklin and at the Sterling mine nearby, sometimes in surprisingly large masses. These sometimes fluoresce an orange-brown or burnt orange under short wave and may be found with fluorescing willemite or calcite. The fluorescence is usually rather weak, but an occasional specimen is found with a rather strong and pronounced orange response. These fluorescent responses may be due to manganese and arsenic, as discussed under svabite.

In an entirely different setting, apatite is commonly found in pegmatites in the form manganapatite. In this mineral, substantial amounts of manganese replace calcium in the crystal structure. Such apatite is usually pale gray or pale green and generally difficult to distinguish from the host feldspar in which it is found, but under the ultraviolet light, a startling transformation takes place. Under short wave, the apatite fluoresces yellow, and under long wave orange-yellow, in sharp contrast to the usually nonfluorescent feldspar. In New England, fluorescing manganapatites are found in the pegmatite dumps in Vermont and New Hampshire. At the Strickland quarry in Connecticut, it is often accompanied by blue or violet fluorescing feldspar or green fluorescing hyalite. Sometimes the three fluorescents – manganapatite, feldspar, and hyalite – are found in one specimen, furnishing an unusual and interesting combination. Yellow-orange fluorescing manganapatite is also found in the pegmatites of Little Switzerland in North Carolina, and Harding, New Mexico. In a different geological setting in some ways similar to a pegmatite, peach or pink-orange fluorescing apatite has been found at the Edison mine in Sussex County, New Jersey. This apatite may be found in blue or red fluorescing feldspar, and may be accompanied by blue and yellow fluorescing fluorite. In yet another setting, apatite is found at Holcomb Valley, California. The material is quite interesting, since the orange-yellow fluorescing apatite is set in a contrasting red fluorescing calcite.

The activator at work in these yellow or orange fluorescing apatites is considered to be fairly reliably established. Experiments with synthetic apatite indicate that manganese is the activator, with a few percent of manganese being sufficient to produce the yellow or yellow-orange response. Lead, or perhaps antimony or tin, must also be present. What is at work here is a dual activator system as with calcite and several other manganese activated minerals. The lead must be present to capture ultraviolet radiation, and the captured energy is transferred to the manganese atoms which then radiate yellow light.

The apatites above are fluorine apatites, including the manganese bearing variety. In contrast, a chlorine apatite from Taylorville, California, fluoresces an intense orange under short wave; the activator is not known.

Additional information on apatite will be found in Part B of this chapter.

Svabite, $Ca_5(AsO_4)_3F$.

Svabite is an arsenic analogue of apatite. In svabite, the arsenic atoms take the place of the phosphorus atoms of apatite. Natural svabite is part of a continuous series with apatite, and specimens are found with varying amounts of both arsenic and phosphorus. Probably the only sources of svabite in the United States are the mines at Franklin and Ogdensburg, New Jersey. Since apatite is also found at these mines, and since the two minerals are indistinguishable by sight, a natural problem of identification presents itself.

It has become the custom to call material found well out in the Franklin marble apatite on the assumption that little or no arsenic is available for its formation, and to call material found in the ore svabite on the assumption that the ore has supplied the required arsenic. The first assumption is probably correct, but the second is in some doubt. Specimens found in the ore probably contain arsenic, but according to mineralogical convention, it is only when the arsenic exceeds the phosphorus that the mineral is to be called svabite. Thus, it is difficult to report on the fluorescence of svabite from Franklin, since much of what is called svabite is more likely an arsenic apatite. Whether svabite or arsenic apatite, these fluoresce brown-orange, burnt orange, or pink-orange under short wave ultraviolet. As a point of comparison, svabite from Sweden fluoresces pink to orange under short wave.

Apatite from the ore body contains a moderate amount of manganese and is thus manganapatite which at other locations fluoresces yellow. Experiments with synthetic manganapatite containing some arsenic in place of phosphorus show that the manganapatite fluorescence is moved to longer wavelength by the presence of arsenic, that is, toward orange or red. It is likely, then, that the orange and pink fluorescences seen in material from Franklin represent a manganapatite fluorescence rendered orange or pink by the presence of some arsenic. It is possible that Franklin specimens which fluoresce a color close to pink are truly svabite, that is, they contain more arsenic than phosphorus. More work needs to be done to relate the fluorescent color of such Franklin material to its chemical makeup, particularly its phosphorus/arsenic ratio.

Additional information on svabite will be found in Part B of this chapter.

Johnbaumite, $Ca_5(AsO_4)_3(OH)$.

Johnbaumite is a very rare hydroxyl analogue of svabite found at Franklin, New Jersey. The fluorescence is reported to be in all respects identical to that of svabite, but weaker.

Pyromorphite, $Pb_5(PO_4)_3Cl$.

The abandoned lead mines near Phoenixville, Pennsylvania, have been recognized as one of the outstanding sources of oxidation zone or supergene mineral specimens. The green, perfectly hexagonal, pyromorphite crystals from these mines are among the best in the world. Rarely, these green pyromorphites will fluoresce an attractive yellow-orange under long wave light.

Additional information on pyromorphite will be found in Part B of this chapter.

Hedyphane, $(Ca,Pb)_5(AsO_4)_3Cl$.

Hedyphane can be considered a calcium rich mimetite. One specimen of Franklin, New Jersey, hedyphane which was examined was white in ordinary light, and fluoresced dull violet under long wave ultraviolet, while others fluoresced a very weak red or pink under long wave. Reports have hedyphane as weak white under both long and short wave ultraviolet. A tan variety was reported to fluoresce a weak orange under both wavelengths. All of this suggests that more study of this mineral is needed.

Additional information will be found in part B of this chapter under hedyphane and mimetite.

Hydrous Acid Phosphates, Divalent Bases

Guerinite, $Ca_5H_2(AsO_4)_4 \cdot 9H_2O$.

A white arsenate from the Sterling mine in Ogdensburg, New Jersey, guerinite has a white response under both short and long wave ultraviolet.

Picropharmacolite, $H_2Ca_4Mg(AsO_4)_4 \cdot 11H_2O$.

Another rare arsenate from the Sterling mine at Ogdensburg, New Jersey, picropharmacolite has a fluorescent response which is white, blue-white, or violet-white, strongest under short wave.

Other Phosphates

Herderite, $CaBe(PO_4)(OH)$.

Herderite, or more exactly hydroxyl-herderite, is found as transparent crystals at the Waisanen mine in Greenwood, Maine. These crystals may fluoresce yellow-white under short wave ultraviolet. The fluorescence appears to originate in a surface alteration of the mineral.

Adamite, $Zn_2(AsO_4)(OH)$.

The spectacular fluorescing adamites from Mapimi, Mexico, are well known, but where are the fluorescent adamites from the United States? They are to be found at Gold Hill, Tooele County, Utah, as small, clear, yellowish crystals on iron stained matrix. The fluorescence is a weak to moderate green under short wave or long wave ultraviolet, with some phosphorescence.

These adamites in no way rival the Mexican material as fluorescent specimens. One study attributes the green fluorescence of adamite from Mapimi to the presence of small amounts of uranium in the form of uranyl. The color of the fluorescence of these adamites supports this view. It is thus likely that uranium is the activator in the adamite from Utah as well.

Tilasite, $CaMg(AsO_4)F$.

Tilasite is another of those mineral rarities which continue to be turned up at the Sterling mine at Ogdensburg, New Jersey. It is reported as whitish crystals up to 4 mm in length, as well as in massive form. The crystals are reported to fluoresce a soft pink-orange or cream-yellow under short wave ultraviolet, with no response under long wave. The massive material is reported not to be a fluorescent.

Variscite, $AlPO_4 \cdot 2H_2O$.

Nodules of variscite from Fairfield, Utah, are found in a delicate and beautiful emerald green color. Sometimes the variscite nodules contain thin veins of a light tan material. Both minerals may fluoresce. Under long wave ultraviolet, the variscite fluoresces grass green. The tan material fluoresces lemon yellow. Under short wave, the response is similar but weaker.

Uralolite, $CaBe_3(PO_4)_2(OH)_2 \cdot 4H_2O$.

Found with other beryllium phosphates at the Dunton Gem mine in Newry, Oxford County, Maine, uralolite fluoresces yellow-green (much like autunite) under short wave and butter yellow under long wave. The fluorescence of uralolite has been suggested as a handy way to distinguish this mineral from the other phosphates found with it. The long wave yellow response will distinguish this mineral from most fluorescent uranium minerals.

Wavellite, $Al_3(PO_4)_2(OH)_3 \cdot 5H_2O$.

Wavellite from Kreamer, Pennsylvania, occurs in spheres to the size of a pea. Broken specimens reveal that these balls are composed of dense, silky white fibers or blades radiating from the center of the sphere. The center portions of the spheres fluoresce bright sky blue – and the outer portions bright yellow – under long wave, weaker under short wave. Both areas phosphoresce strongly. Wavellite from Hellertown, Pennsylvania, consists of randomly oriented sections of silky white fibers in cavities of the host rock. Fluorescence is butter yellow, strongest under long wave. The very well-known green wavellites from Arkansas are usually nonfluorescent, the green color suggesting the presence of excessive iron. However, a scarce few of these specimens are weakly fluorescent and phosphorescent, particularly the lightly colored ones.

Fluellite, $Al_2(PO_4)F_2(OH) \cdot 7H_2O$.

Fluellite can be considered the rarest United States mineral reported here since the report is based on one specimen seen from this location. The source is the pyrophyllite mine at Staley in Randolph County, North Carolina. The mineral was found in association with pyrophyllite. The fluorescence is reported as creamy white under long wave length, which may make it hard to find on the basis of fluorescence since pyrophyllite often fluoresces in this color.

Cahnite, $Ca_2B(AsO_4)(OH)_4$.

A rare mineral from Franklin, New Jersey, a specimen of cahnite in the Gerstmann Museum at Franklin shows a very weak butter yellow response under long wave and no response under short wave.

Reports of the fluorescence of a few other specimens indicate a somewhat stronger fluorescence, with response under both short and long wave ultraviolet.

SILICATES

Nesosilicates

Willemite, Zn_2SiO_4.

The mine at Franklin, New Jersey, is now closed, and the shafts and pits that yielded so many famous and unique minerals are now water filled. The nearby Sterling mine at Ogdensburg, working a similar ore body, is still in operation, producing hand size mineral specimens only infrequently since the ore is now crushed below ground.

At almost all major zinc deposits throughout the world, sphalerite is the main zinc ore. At Franklin and Ogdensburg, however, it is primarily willemite. This willemite occurs in the Franklin marble in association with one or another of the many minerals which occur in these mines. It is found as massive material, sometimes as colorful crystals locked in marble (calcite), and very rarely as free crystals in vugs. Almost all of the willemite is fluorescent and most phosphorescent, usually brilliant green under short wave and weaker under long wave. In the early decades of this century, willemite ore was identified under crude iron arc ultraviolet lamps in what was one of the earliest industrial applications of ultraviolet fluorescence.

The combination of willemite and calcite is the most commonly found fluorescent combination at Franklin and Ogdensburg. Under short wave ultraviolet, the combination provides a spectacular display of fluorescence, the fiery red of the calcite complementing almost perfectly the brilliant green of the willemite. This combination may be the most attractive of all fluorescent mineral specimens.

Both the willemite and the calcite owe their fluorescence to the same activator – manganese. The source of the manganese is not hard to find. Manganese is intrinsic to the willemite, displacing zinc in the crystal lattice to the degree of between 1 and 13%. Maximum fluorescence in willemite is produced at a manganese concentration of about 1%. Manganese is thus not a trace activator, but one which may be present in relatively high concentration. At the optimum concentration of activator, willemite is possibly the brightest of fluorescents, and it is possible to read under the light given off. This is particularly true of the white, powdery willemite which occurs as a vein filling. This may be a secondary deposition, freed of iron which is ordinarily harmful to fluorescence, and with manganese reduced to the optimum concentration. These white willemites in particular have an intense phosphorescence. Arsenic as an impurity is known to enhance phosphorescence in willemite, and since this element occurs widely at both Franklin and Ogdensburg, we may attribute the phosphorescence of willemite to arsenic.

Not as well known is a willemite which occurs rarely at Ogdensburg as a bright yellow or yellow-orange fluorescent under short wave ultraviolet, with an enduring yellow phosphorescence. Even less widely known is a similarly fluorescing willemite from the iron mine at Andover about ten miles away, not ordinarily thought of as a source of willemite. This willemite usually occurs as a small mass of sugary white or tan crystals in calcite. The cause of this unusual yellow-orange fluorescence is open to some debate. Synthetic willemite has been prepared under conditions of fast formation which produces a disordered crystal lattice resulting in yellow fluorescence or, with complete disorder, red. Whether this process occurs in nature is not clear, but it seems a doubtful explanation for the above fluorescence, as the material is visibly crystalline. However, copper as an activator in willemite also produces a yellow fluorescence, and the yellow fluorescence and phosphorescence of willemites from Ogdensburg and Andover may be attributed to copper activator or perhaps copper in combination with lead.

The avid Franklin collector will hopefully not be offended to learn that fluorescent willemite, or even fluorescent combinations of willemite and calcite, can be found elsewhere in the United States. Willemite occurs in metal deposits over a broad geographic belt covering western New Mexico, Arizona, Utah, Nevada, and eastern California, frequently with red fluorescent calcite. Green fluorescent willemite occurs at Socorro, New Mexico; Casa Grande, Wickenburg, Moctezuma Canyon, Bisbee, and the Red Cloud mine, Arizona; Holcomb Valley, California; and numerous other locations. In the East, green fluorescing willemite has been found with orange fluorescing sphalerite at the St. Joe No. 3 mine at Balmat, New York.

A particularly unusual willemite has been found at the Mammoth mine at Tiger in Arizona, fluorescing greenish white under short wave and yellow-orange or burnt orange under long wave, suggesting the possibility of a mixture of copper and manganese activators at work. These attractive crystal clusters are found with green dioptase crystals and orange wulfenite crystals to make an attractive and interesting specimen in ordinary light. This is one of those exceptional and intriguing specimens which fluoresce in different colors under short and long wave.

Eucryptite, $LiAlSiO_4$.

Eucryptite comes from Stafford, New Hampshire, and appears waxy and translucent under ordinary light, and white or slightly gray or gray-brown in color. Under short wave ultraviolet, this drab color is replaced by an attractive bright deep red or cerise fluorescence over part or all of the specimen. The portion which does not fluoresce deep red may fluoresce blue-white or gray. It is reported that when the material has been exposed to daylight for some time, the red fluorescent capability disappears. It can be regained if a fresh surface is broken open. Under long wave, only the blue-white or gray fluorescence is sometimes seen. Quick removal of the short wave lamp reveals a brief red phosphorescent flash.

Similar fluorescent eucryptite is reported from the lithium pegmatites at Custer, South Dakota; the Foote mine at Kings Mountain, North Carolina; and the Harding mine in Taos County, New Mexico. A particularly attractive eucryptite found as small crystals on matrix, is reported to originate from the Midnite Owl mine near Wickenburg, Arizona, fluorescing a vivid red, almost magenta, under short wave ultraviolet.

Forsterite, $(Mg,Fe)_2SiO_4$.

Forsterite from the Crestmore quarry in Riverside, California, consists of white, chalky masses fluorescing white or tan, best under long wave. Based on the white color, the iron content of this forsterite is probably extremely low, a condition which is probably necessary for fluorescence. Forsterite from Limekiln Canyon, Pacoima, California is reported to fluoresce blue.

Zircon, $ZrSiO_4$.

Zircon is frequently found in igneous rocks, and its occurrence is widespread, though rarely in large or showy enough specimens to be attractive as a collector's mineral. However, zircon is often a brilliant fluorescent. The fluorescent color seems invariably to be an intense yellow-orange, sometimes most intense under long wave ultraviolet light and in other specimens more intense under short wave.

Fluorescent zircon occurs in pegmatites and dikes in the uplands of northern New Jersey. One such occurrence is with fluorescent sodalite in nepheline at Beemerville, New Jersey. Downstream in southern New Jersey, fluorescent zircon has ended as a constituent of ancient sand deposits from which it is mined. It is also found sparingly in the sands of all New Jersey beaches as small clear grains in washes of black sand which are left behind as waves retreat from the beach. The black sand is ilmenite with which the zircon is mixed, and these can be separated from silica sand by panning.

Zircon occurs as pink crystals in a pegmatite at Pacoima, California; these crystals fluoresce yellow. It is reported as a yellow-orange fluorescent from Statesville, North Carolina; the Ruggles mine in Grafton, New Hampshire; Colorado Springs, Colorado; the Crestmore quarry in Riverside, California; and various sand deposits in Idaho and Oregon.

At Franklin, New Jersey, yellow fluorescing zircon occurs in a dike cutting the ore body. It is considered a rare Franklin fluorescent. It is also found in the so-called pegmatites at Lime Crest quarry in Sparta, New Jersey.

The cause of fluorescence in zircon is not fully established. Some attribute it to the presence of rare earths, while others suggest that the fluorescence is due to the element hafnium, which always accompanies zircon in small amounts.

Norbergite (and Chondrodite), $Mg_3SiO_4(F,OH)_2$.

Norbergite and chondrodite are two very closely related minerals which are difficult to distinguish from one another. It used to be taken as a rule among collectors that the colorless or yellow material is norbergite, while the darker or brown material is chondrodite. It is doubtful if there is any basis in fact for this assumption, so that it is not clear what we should call the fluorescent specimens when found.

The mineral is usually found as yellow, orange, or brown blebs or masses, from one-quarter of an inch to occasionally several inches in size in the Franklin marble in quarries at Sparta and Franklin, New Jersey. It is also found in New York at Brewster, Newcomb, and in the St. Joe mine at Balmat. In some locations, it is white or colorless and thus almost indistinguishable from the white calcite in which it is embedded. Under short wave ultraviolet, the lighter material fluoresces a fairly bright lemon yellow; the darker material, a less bright orange-yellow. The response is only occasional and weak under long wave.

The most attractive specimens are those with a large number of bright yellow fluorescing blebs or those in which yellow fluorescing material is intimately mixed with diopside which fluoresces sky blue, thus providing a very attractive two-color fluorescent combination.

Hodgkinsonite, $MnZn_2SiO_5 \cdot H_2O$.

A rare Franklin mineral consisting of small red-purple crystals, hodgkinsonite sometimes is found in other shades of red. This mineral sometimes fluoresces dull dark purple under long wave; other reports describe the fluorescent color as red.

Datolite, $CaBSiO_4(OH)$.

Datolite is an occasional fluorescent. Some crystal specimens from the trap rocks in or near Paterson, New Jersey, sometimes fluoresce a fairly bright yellow-green under short wave, with weak yellow, gray, or no response under long wave. In clusters of datolite crystals, one may fluoresce while another nearby one may not. Also, the response seems to derive from a portion of the crystal rather than the entire crystal in most cases. Masses of glassy gray datolite are found in the Crestmore quarry, Riverside, California; these fluoresce an intense green, possibly due to a surface alteration to hyalite.

Titanite (Sphene), $CaTiSiO_5$.

Fluorescent titanite has been found in the Franklin marble at Rudeville, New Jersey. The fluorescent color is a weak brown under short wave ultraviolet. The tin analogue of titanite (see malayaite in Part B of this chapter) is a bright fluorescent so that the fluorescence of titanite is not totally surprising. What is surprising is the apparent rarity of fluorescing titanite. Further search may show that it is more common than is now apparent.

Thaumasite, $Ca_3Si(OH)_6(CO_3)(SO_4) \cdot 12H_2O$.

Thaumasite is one of the reluctant and undistinguished fluorescents from the trap rocks around Paterson, New Jersey. The response, on the few occasions when it is seen, is a dull and uneven white under long wave. The response of thaumasite from Crestmore quarry in Riverside, California, is similar to the above.

Harkerite, $Ca_{24}Mg_8Al_2(SiO_4)_8(BO_3)_6(CO_3)_{10} \cdot 2H_2O$.

Harkerite has been identified among the minerals of the Crestmore quarry in Riverside, California. It thus adds to the list of interesting minerals recognized from this renowned location. The fluorescence is weak dark red under long wave ultraviolet. Possibly some material identified as merwinite may really be harkerite.

Roeblingite, $Pb_2Ca_7Si_6O_{14}(OH)_{10}(SO_4)_2$.

Roeblingite is a Franklin mineral, chalky and white in appearance. The fluorescence of roeblingite is denied by many Franklin specialists, but what can be said with certainty is that at least a few specimens labeled roeblingite fluoresce a clear, full red under short wave. The response is similar but weaker under long wave. That such specimens are roeblingite is usually not disputed. What is said, however, is that the fluorescence must be due to admixed calcite. The support for this assertion is that hydrochloric acid, which dissolves calcite, destroys the fluorescence. This argument is not wholly convincing since the roeblingite crystal structure or a legitimate activator may be changed by this strong treatment.

Dumortierite, $Al_7(BO_3)(SiO_4)_3O_3$.

An interesting mineral from San Diego County, California, dumortierite is found as veins of bladed or fibrous crystals, maroon or purple in color, sandwiched in a matrix of a milky white mineral. The dumortierite fluoresces a light sky blue under short wave ultraviolet, while the white material fluoresces light yellow under long wave ultraviolet, providing an attractive two-color fluorescent. Reddish, bladed dumortierite is found at Oreana, Nevada, fluorescing white under short wave ultraviolet.

Howlite, $Ca_2B_5SiO_9(OH)_5$.

Howlite from the Sterling mine in Tick Canyon, Los Angeles County, California, fluoresces bright sky blue. The response is particularly strong under long wave ultraviolet.

Foshagite, $Ca_4Si_3O_9(OH)_2$.

Foshagite, a hydrated calcium silicate, is reported to fluoresce yellow-white or blue-white under short wave ultraviolet. The source is the Crestmore quarry in Riverside, California.

Sorosilicates

Hardystonite, $Ca_2ZnSi_2O_7$.

Hardystonite is another of the famous fluorescents from Franklin, New Jersey, usually attributed to the Parker shaft. From the frequency with which it is found on the shelves of Franklin dealers, it may be judged to be the third most common of the Franklin fluorescents, following calcite and willemite. While it is fairly common at Franklin and while the fluorescence is usually not outstandingly brilliant, it is sometimes outstandingly beautiful. Under short wave the fluorescence is deep dark violet-blue, almost dark purple. It frequently fluoresces under long wave ultraviolet, usually an intense violet-blue. Hardystonite is usually found in specimens several inches across, and the large size of the fluorescing area adds to the attractiveness of the fluorescence. It is rarely if ever found as crystals. Rather, it is usually found as a massive with a glassy to resinous luster, and a white, gray-white, or slightly pink-tan color. However, in one specimen seen, there appear to be hardystonite crystals, though it is possible that these are hardystonite replacing a previous mineral. The crystals are over one inch long and one-half inch thick, oriented randomly in a willemite matrix. The hardystonite fluoresces violet-blue, best under long wave, and the willemite fluoresces its usual green.

Experiments have been performed with synthetic hardystonite activated with small amounts of lead. Such hardystonite fluoresces intensely, with peak output in the long wave ultraviolet but with some response extending into the visible violet and blue portions of the light spectrum. This suggests the possibility that lead may be the activator in natural hardystonite.

Hardystonite is frequently found with red fluorescing calcite, sometimes as small blebs in calcite, and the two fluorescent colors, purple and red, create a vivid and very attractive contrast. Hardystonite is also often found with fluorescing clinohedrite, or with fluorescent willemite or esperite. Rarely, the collector may come across a specimen containing all of these minerals in close association, and the fluorescent combination of violet-blue, red, orange, green, and yellow is unmatchable in any other specimen from anywhere in the world.

Barylite, $BaBe_2Si_2O_7$.

Barylite, a very rare Franklin mineral, has long been listed as a blue or violet fluorescent; more recently, this claim has been the subject of some dispute. The original claim for the fluorescence of barylite was due to Charles Palache, the dean of Franklin mineralogists. His investigations of fluorescence were performed using the iron arc as a source of ultraviolet. This pioneer machine was instrumental in discovering many fluorescent minerals, but one of its defects is that it emits substantial visible light. As this light is somewhat violet in color, claims made for violet or blue fluorescence are unreliable to a degree, since what is seen may simply be a reflection of visible light.

An examination of barylite made with modern ultraviolet light sources may reveal no response under either short or long wave. An examination of a crystal of barylite from Langban, Sweden, showed no fluorescence.

However, one specimen labeled barylite in the museum at Paterson, New Jersey, consisting of a cluster of crystals, showed a positive and reasonably certain response under a modern, well-filtered short wave light. The fluorescence was a weak lavender, with some sections surprisingly fluorescing red. Close examination in ordinary light showed no distinction between the lavender and the red fluorescing portion of the crystal. Since a weak lavender response under an ultraviolet light can be due to light leakage, a sheet of clear plastic was interposed between the light and the specimen. Such a sheet will stop short wave ultraviolet but not visible light. The result was that the lavender response was extinguished, indicating a true fluorescence. A barylite in the Gerstmann collection, Franklin, New Jersey, revealed a fluorescence in all respects identical to that described above, as does a specimen in the author's collection. This specimen definitely fluoresces lavender under short wave, with spots of red. To confirm the mineral identification, a microanalysis was made. The results showed barium, beryllium, and silicon in the amounts to be expected of barylite. Lead was also present to about 1%. Barylite should, therefore, be considered a fluorescent mineral.

Hemimorphite, $Zn_4Si_2O_7(OH)_2 \cdot H_2O$.

Hemimorphite is usually rather undistinguishedsas a fluorescent. It is found as white crystal druses at Ogdensburg, New Jersey, fluorescing a dull, weak white or gray-white under both long and short wave. A similar response is reported in material from Superior, Arizona. However, one specimen from Ogdensburg fluoresced grass green under short wave and yellow under long wave. Whether this is due to the hemimorphite itself or some response of the matrix beneath is not clear. Hemimorphite from Joplin, Missouri, is reported to fluoresce blue-white, light yellow, or light green.

Clinohedrite, $CaZnSiO_3(OH)_2$.

Another of those fluorescents taken out through the Parker shaft at Franklin, clinohedrite like many fluorescents from this famous shaft may be unique to this locality. The fluorescence is a medium to strong orange under short wave ultraviolet. Occasionally, a weaker response to long wave will be seen. The fluorescent responses of clinohedrite, pectolite, and wollastonite from Franklin are substantially tially the same, and it is difficult to tell these minerals apart on the basis of their fluorescence. A more reliable identification can often be made on the basis of daylight appearance and mineral association. In daylight, clinohedrite is occasionally white, but more often it has a pink color with a shading toward purple or brown. Orange fluorescing crystals of clinohedrite are known and are quite rare.

Clinohedrite is an almost constant companion of hardystonite, and few hardystonite specimens will be seen without some orange fluorescing clinohedrite. Usually, such clinohedrite covers a surface of the hardystonite as a film, and in such cases, little of the daylight pink color will be evident.

Clinohedrite is also found with fluorescent esperite and sometimes with the rare fluorescent margarosanite, while the common fluorescent willemite is usually close by.

Cyclosilicates

Benitoite, $BaTiSi_3O_9$.

The strictly American gemstone benitoite occurs as deep violet-blue, triangular crystals usually about three-eighths of an inch across, occasionally an inch or more across. The setting of the crystals in a bed of snow white natrolite on serpentine makes such specimens outstandingly attractive.

Benitoite fluoresces a bright deep sky blue under short wave ultraviolet light and is one of the most beautiful of fluorescent minerals. At the tips of many benitoite crystals, the blue daylight color of the crystal sometimes gives way to white. These white sections frequently fluoresce pink-red, sometimes fairly brightly, under long wave ultraviolet.

Benitoite was first discovered in 1907 near the mining town of Coalinga in San Benito County, California, and until recent years the mineral was known only from this one site. It is now known from a few other localities in the United States, including the walstromite site in Fresno County, California, but the original mine remains the prime source.

A closely related fluorescent mineral, pabstite, comes from nearby Santa Cruz County, and another related fluorescent mineral, fresnoite, comes from elsewhere in San Benito County.

Pabstite, $Ba(Sn,Ti)Si_3O_9$.

The rare mineral pabstite is found in Santa Cruz, California, not a great distance from the source of the fluorescent mineral benitoite. This is particularly interesting, since pabstite is a tin analogue of benitoite. Benitoite is a barium titanium silicate, while pabstite is virtually identical except that tin atoms substitute for most of the titanium atoms. Fluorescence is indistinguishable from that of benitoite – a deep, intense sky blue under short wave ultraviolet.

Pabstite is a colorless or gray material almost indistinguishable from the equally gray stone in which it is found. A silvery sulfide is an accompaniment in some specimens. This may be stannite which is reported to be present, and it may be the source of the tin which has gone into the composition of pabstite.

Axinite, $Ca_2(Mn,Fe)Al_2BSi_4O_{15}(OH)$.

Axinite is another of those scarce but beautiful fluorescents which came out of the Parker shaft at Franklin, New Jersey, often in combination with fluorescent willemite, hardystonite, clinohedrite, and margarosanite. Axinite fluoresces a bright cherry red under short wave ultraviolet and, usually but not always, a weaker red under long wave.

The variety which fluoresces is manganaxinite, that is, axinite with substantial substitution of manganese for calcium. Not all of the manganaxinite fluoresces, and of two specimens that may appear to be identical in ordinary light, one may fluoresce and the other may not. Ferroaxinite from the Sterling mine at nearby Ogdensburg does not fluoresce.

It is probably manganese which is responsible for the red axinite fluorescence, since manganese is a red fluorescent activator in a wide variety of minerals. Similarly, the iron in ferroaxinite is probably responsible for the absence of fluorescence in this mineral and in axinites from many other locations.

While crystals of fluorescent manganaxinite are known, the collector will seldom see them. The axinite at Franklin generally occurs in masses, usually a few inches across. The daylight color is tan to yellow, and this color, together with high hardness and the absence of a phosphorescent flash, helps to distinguish axinite from calcite, which has a very similar fluorescence.

Beryl, $Be_3Al_2Si_6O_{18}$.

Beryl seems to fluoresce only very infrequently. A vivid and clear fluorescent response was seen in a clear, colorless beryl from Mount Apatite, West Minot, Maine. The fluorescence was bright yellow-orange under both long and short wave ultraviolet. A colorless white crystal of beryl from Pend Oreille County, Washington, fluoresces identically to the above. Similar fluorescence is reported in beryl from East Wakefield, New Hampshire, while a weak yellow response, under long wave only, is reported in beryl from Newry, Maine.

Additional information on beryl will be found in Part B of this chapter.

Tourmaline (Uvite), $(Ca,Na)(Mg,Fe)_3Al_5Mg(BO_3)_3Si_6O_{18}(OH,F)_4$.

The uvite variety of tourmaline is found in the Franklin marble of New Jersey and New York. This variety is comparatively iron free, colored faintly brown-green, and is seen as crystals foreshortened along the main axis, sometimes quite stubby. Fluorescence under short wave is yellow to orange-yellow, with little or no response under long wave.

Tourmaline (Elbaite), $Na(Li,Al)_3Al_6(BO_3)_3Si_6O_{18}(OH)_4$.

Large crystals of "watermelon" tourmaline have been found in feldspar quarries at Newry, Maine. These crystals are pink at the center and green on the outside, thus the name watermelon. It is surprising to discover that these crystals are fluorescent. The fluorescence is an attractive blue or blue-white under short wave ultraviolet and appears in the pink portion of the crystal.

Inosilicates

Diopside, $CaMgSi_2O_6$.

Diopside occurs frequently in the Franklin marble from Sparta and Franklin, New Jersey, northward into New York. Frequently enough it is fluorescent in a pleasing sky blue under short wave. Quite often, the diopside is found with norbergite which fluoresces bright yellow, providing a very attractive blue and orange fluorescent combination. Similar fluorescent diopside is found at Newcomb, New York, and its fluorescence is an even richer sky blue. Here, it is often found with phlogopite which fluoresces a good yellow, and such specimens are quite similar in appearance and attractiveness to the diopside and norbergite mentioned above. Diopside is also found as very large masses in a pegmatite in northern New Jersey, fluorescing blue-white, less blue than the diopsides mentioned earlier. Under long wave, these diopsides fluoresce a not very attractive pale yellow.

Spodumene and Kunzite, $LiAlSi_2O_6$.

Specimens of tan spodumene from the Harding mine in Dixon, New Mexico, are found with muscovite of an unusual wine color in ordinary light, furnishing interesting specimens. The spodumene fluoresces yellow under ultraviolet, best under long wave. Occasionally, spodumene from the Foote mine in Kings Mountain, North Carolina, shows traces of this response, and a similar response is reported in spodumene from Branchville, Connecticut, and from the Etta tin mine in Pennington County, South Dakota. Other reports indicate a pink fluorescence in spodumene from these three locations. Spodumene sometimes also shows a weak blue-gray fluorescent response.

Since pink fluorescence is seen in eucryptite, it is possible that an alteration to eucryptite is present in those spodumene samples that show pink fluorescence. On the other hand, that response may simply be a subdued version of the more vivid response seen in kunzite.

Kunzite is the lilac or pink gem variety of spodumene. Colorless, clear spodumene is also sometimes referred to as kunzite. The small pegmatite workings in the hills near Pala, California, yield not only pink and red gem tourmalines but occasionally morganite and kunzite. This kunzite is an interesting fluorescent. Its response is pink under long wave, and pink-purple under short wave ultraviolet. The mineral phosphoresces an interesting orange-pink, strongest after short wave excitation. It appears that the clear, colorless kunzites from this locality are the strongest fluorescents and phosphorescents.

PLATE 1

1. Sphalerite fl. deep pink LW, willemite fl. green SW. Sterling mine, Ogdensburg, NJ.

2. Sphalerite fl. peach LW, willemite fl. green SW. Sterling mine, Ogdensburg, NJ.

3. Sphalerite fl. blue LW, willemite fl. green SW. Sterling mine, Ogdensburg, NJ.

4. Sphalerite fl. orange, calcite fl. red, SW. Queen mine, Bisbee, Arizona.

PLATE 2

HALIDES

5. Fluorite fl. blue LW. Weardale, Durham, England.

6. Fluorite fl. butter yellow, LW. Clay Center, Ohio.

7. Calomel fl. orange LW. Terlingua, Texas.

8. Halite fl. red SW. Salton Sea, California.

PLATE 3

9. Fluorite fl. butter yellow LW. Elizabethtown, Illinois.

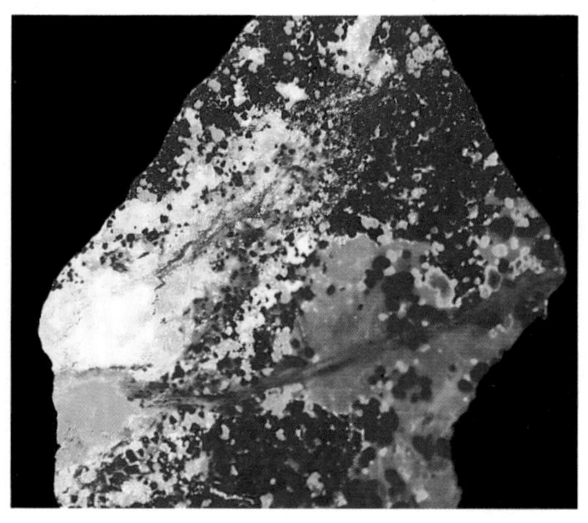

10. Zincite fl. butter yellow, willemite fl. green LW. Sterling mine, Ogdensburg, NJ.

11. Brucite fl. blue-white LW. Cedar Hill quarry, Texas, Pennsylvania.

12. Ruby (corundum) fl. red LW. Mysore, India.

PLATE 4

CARBONATES

13. Aragonite fl. yellow-white, LW. Sterling mine, Ogdensburg, NJ.

14. Calcite fl. yellow-white LW. Bound Brook, NJ.

15. Calcite fl. blue SW. Terlingua, Texas.

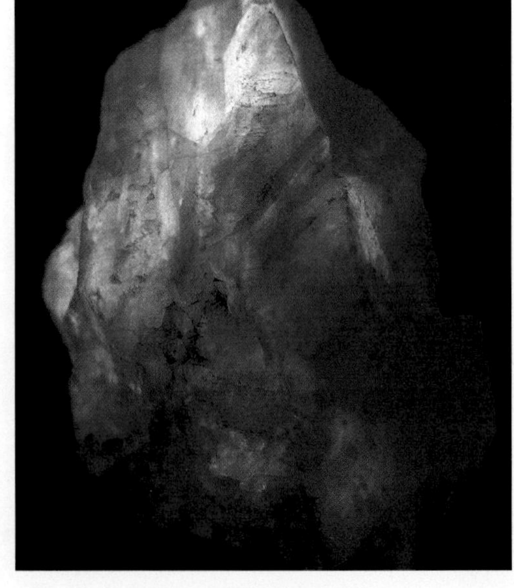

16. Calcite fl. pink LW. Terlingua, Texas.

PLATE 5

CARBONATES

17. Strontianite fl. white LW. Minerva mine, Hardin County, Illinois.

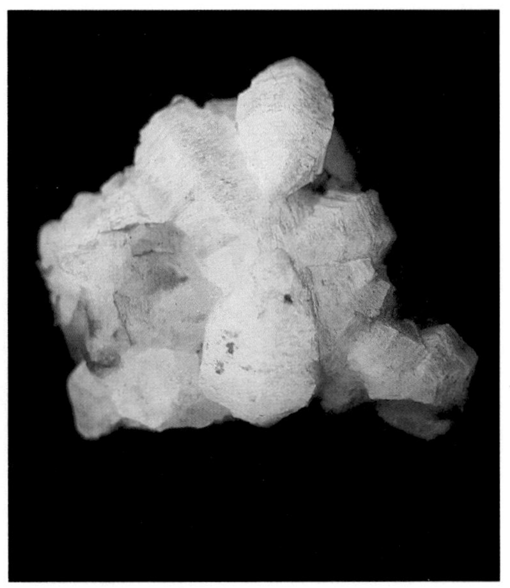

18. Witherite fl. blue-white LW. Roseclare, Illinois.

19. Calcite fl. yellow LW. Joplin, Missouri.

20. Aragonite fl. pink LW. Sicily.

PLATE 6

CARBONATES
BORATES

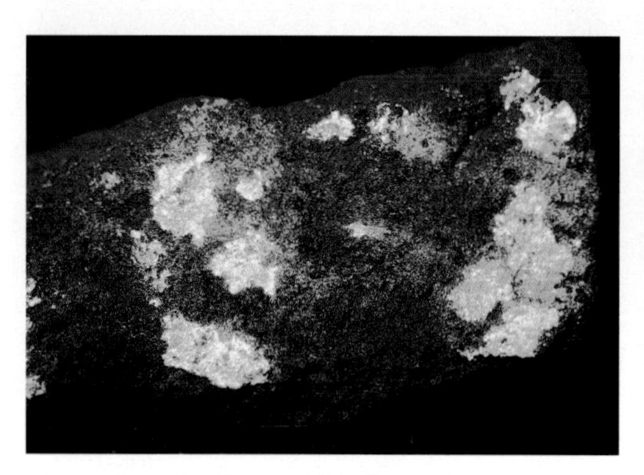

21. Andersonite fl. green SW. Colorado.

22. Cerussite fl. yellow LW. Morrocco.

23. Calcite fl. red, willemite fl. green SW. Sterling mine, Ogdensburg, NJ.

24. Colemanite fl. yellow-white. Death Valley, California.

SULFATES

PLATE 7

25. Celestite fl. blue-white. LW. Sicily.

26. Barite fl. white LW. Minerva mine, Hardin County, Illinois.

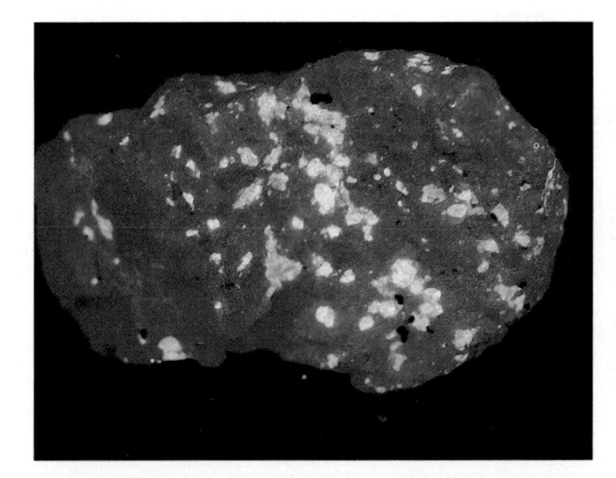

27. Barite fl. yellow, calcite fl. red SW. Franklin, New Jersey.

28. Selenite fl. blue-white and orange LW. Baja California, Mexico.

TUNGSTATES
PHOSPHATES

PLATE 8

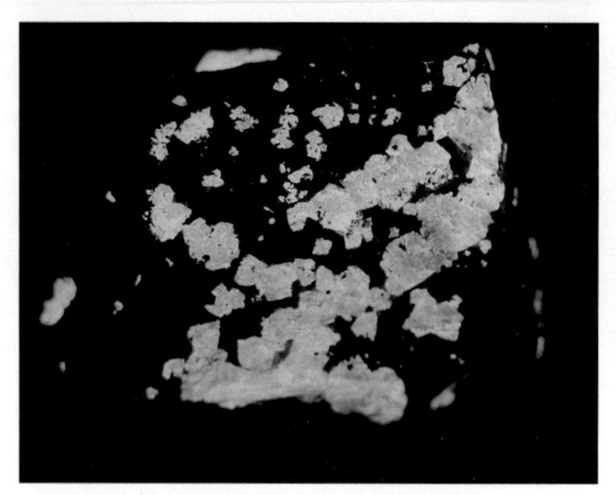

29. Scheelite fl. blue SW. Reedley, California.

30. Scheelite fl. blue, calcite fl. red SW. Oracle, Arizona.

31. Autunite fl. green SW. Daybreak mine, Spokan County, Washington.

32. Apatite fl. blue-grey, clacite fl. red. SW. Bancroft, Canada.

PLATE 9

PHOSPHATES

33. Adamite fl. green, calcite fl. blue-white LW. Mapimi, Mexico.

34. Pyromorphite fl. yellow LW. Ems, Germany.

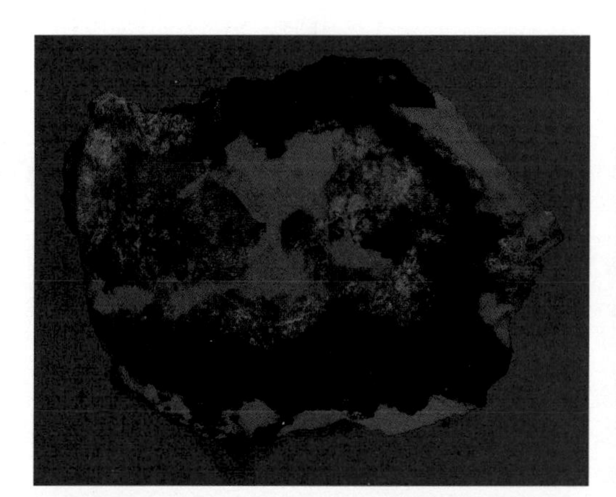

35. Svabite fl. orange, calcite fl. red SW. Franklin, New Jersey.

36. Apatite fl. yellow-orange, calcite fl. red SW. Holcomb Valley, California.

PLATE 10

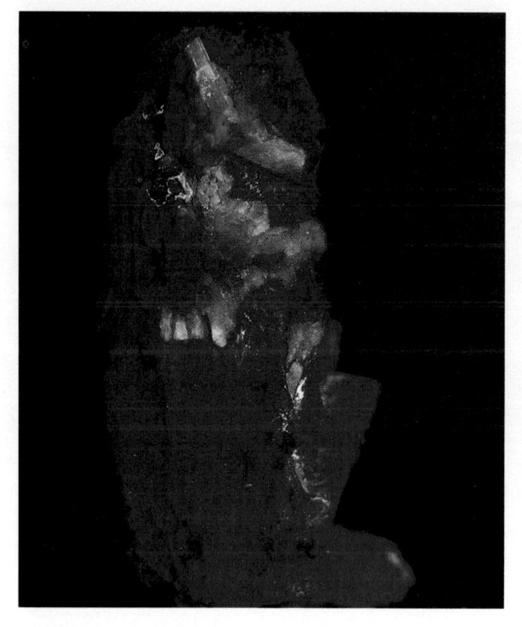

37. Luecophanite fl. magenta-pink LW. Norway.

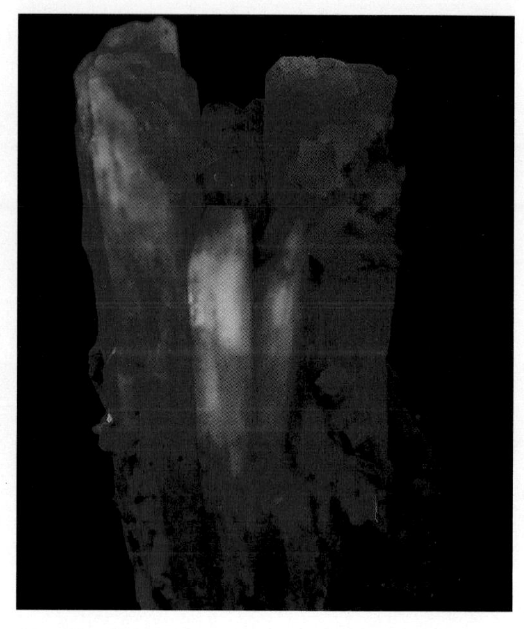

38. Danburite fl. blue, calcite fl. red SW. San Luis Potosi, Mexico.

39. Benitoite fl. blue SW. San Benito County, California.

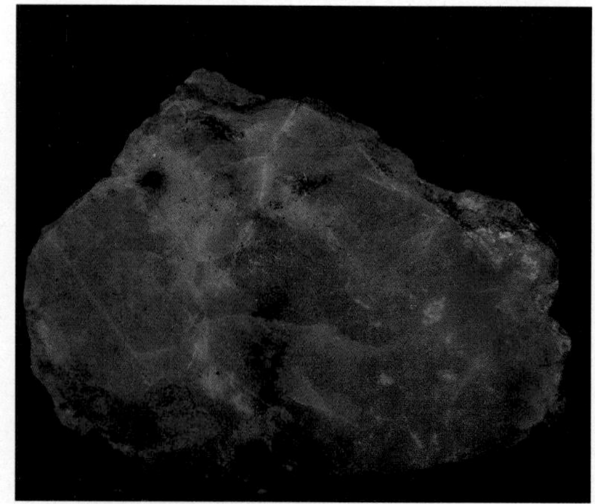

40. Hackmanite fl. orange, LW. Bancroft, Canada.

PLATE 11

SILICATES

41. Wollastonite fl. blue-white LW. Harquahala Mountains, Arizona.

42. Agrellite fl. magenta-pink SW. Quebec, Canada.

43. Zircon crystal fl. yellow SW. Goiaz, Brazil. Zircon sand fl. yellow SW. New Jersey.

44. Chondrodite fl. yellow SW. Franklin, New Jersey.

PLATE 12

45. Tirodite fl. deep red LW, talc fl. butter yellow SW. Talcville, New York.

46. Scapolite (wernerite) fl. yellow LW, diopside fl. blue SW. Ontario, Canada.

47. Tugtupite fl. deep orange LW. Greenland.

48. Tugtupite fl. deep red SW. Greenland.

PLATE 13

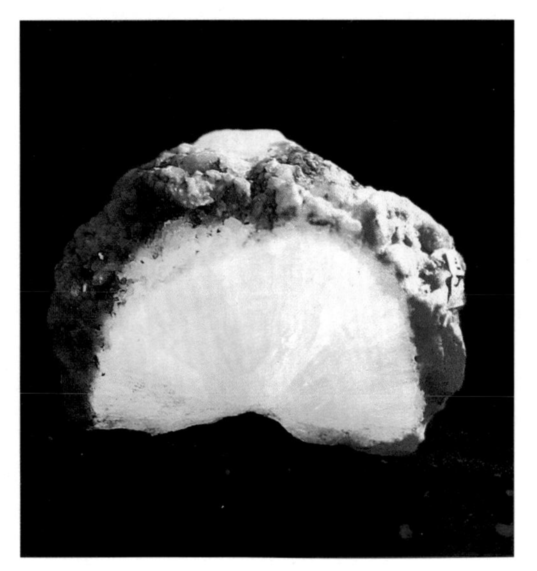

49. Quartz fl. yellow SW. Pisa, Italy.

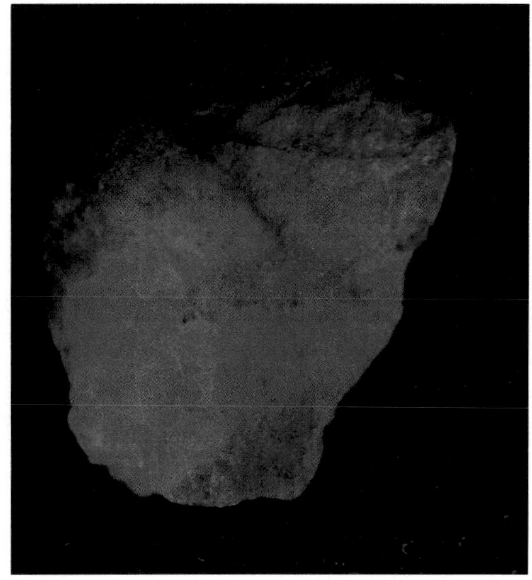

50. Eucryptite fl. red SW. Rhodesia.

51. Chondrodite fl. yellow, dioside fl. blue SW. Franklin, New Jersey.

52. Willemite crystals fl. green, calcite fluorescent blue-white LW. Sterling mine, Ogdensburg, New Jersey.

PLATE 14

SILICATES

53. Esperite fl. yellow, willemite fl. green, calcite fl. red SW. Franklin, New Jersey.

54. Wollastonite fl. orange, calcite fl. red, barite fl. blue-grey SW. Franklin, New Jersey.

55. Margarosanite fl. blue, clinohedrite fl. orange, willemite fl. green, esperite fl. yellow SW. Franklin, New Jersey.

56. Wollastonite fl. orange, willemite green, calcite fl. red SW. Franklin, New Jersey.

PLATE 15

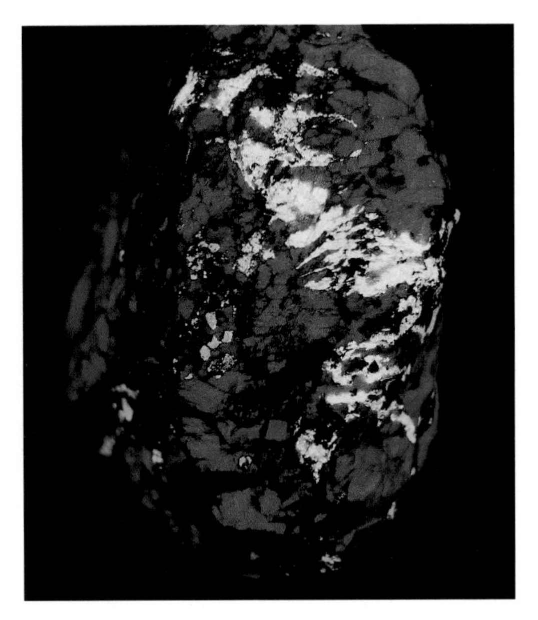

57. Esperite fl. yellow, willemite fl. green, calcite fl. red SW. Franklin, New Jersey.

58. Margarosanite fl. blue, willemite fl. green, clinohedrite fl. orange SW. Franklin, New Jersey.

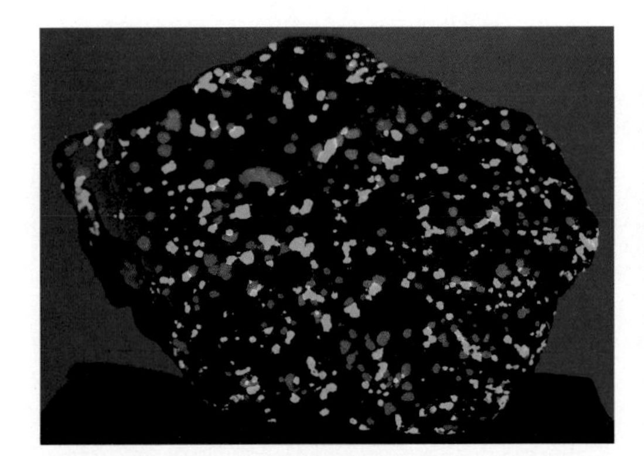

59. "Polka dot ore". Willemite fl. green, calcite fl. red SW in non fluorescing schefferite. Franklin, New Jersey.

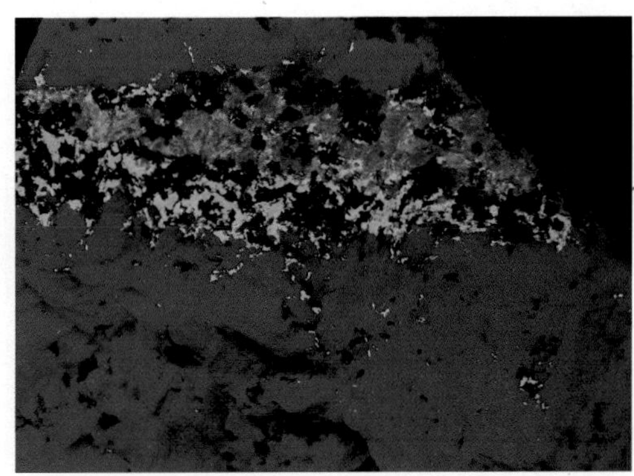

60. Margarosanite fl. blue, willemite fl. green, calcite fl. red SW. Franklin, New Jersey.

PLATE 16

SILICATES
HYDROCARBONS

61. Experite fl. yellow, hardystonite fl. purple, calcite fl. red SW. Franklin, New Jersey.

62. Axinite fl. red, margarosanite fl. blue, willemite fl. green SW. Franklin, New Jersey.

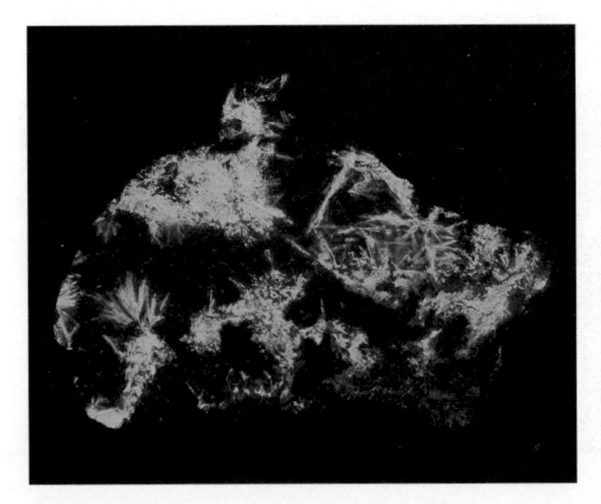

63. Karpatite fl. blue SW. Idria, California.

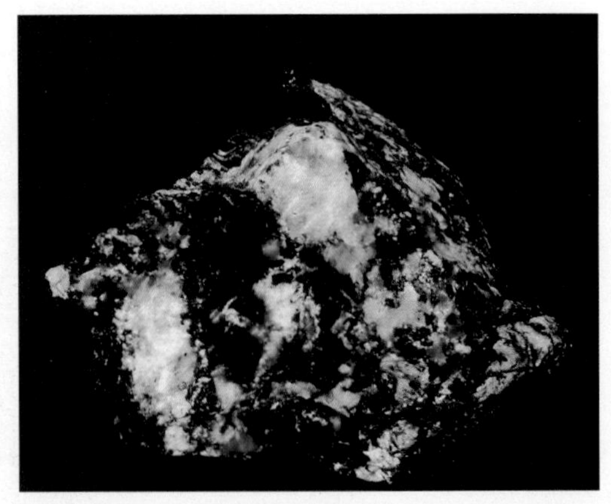

64. Idrialite fl. green, yellow LW. Skaggs Springs, California.

Ultraviolet fluorescence in kunzite has been known since the massive fluorescence survey of Kunz and Baskerville reported in 1903, and it was their observation that only the colorless kunzites responded to ultraviolet. This view is not supported by facts known today, since pink kunzite also fluoresces.

Additional information on kunzite will be found in Part B of this chapter.

Edenite, $NaCa_2Mg_5(AlSi_7)O_{22}(F,OH)_2$.

A member of the amphibole group, edenite is found sparingly at the Goosebury dump, at the Nichols quarry, Franklin, and at Lime Crest quarry, New Jersey. As a fluorescent, it was once considered rare. In daylight, this mineral is dark or olive green and glassy, occurring in crystals or in masses an inch or so in size in the Franklin marble. It is often closely accompanied by norbergite (or is it chondrodite?) and fluorite. These are fluorine bearing minerals and tend to confirm the associated green mineral as a fluorine bearing edenite.

The fluorescence is a weak to bright pale greenish blue under short wave ultraviolet. A yellow fluorescence has also been reported. When found with phlogopite, a nice two-color fluorescent results.

Tremolite, $Ca_2Mg_5Si_8O_{22}(OH)_2$.

Tremolite from the mines at Balmat and Talcville, New York, provides the most attractive fluorescent specimens of this widely occuring mineral. Under both long and short wave, the fluorescence is a tangerine orange. Sometimes also, tremolite from the dumps at these locations will fluoresce pink-orange, yellow-orange, yellow, or occasionally white. However, orange is the most common color. As this mineral may be found in large, uniformly fluorescing pieces, it makes a most attractive fluorescent. This fluorescing tremolite is sometimes found with yellow fluorescing talc and with an attractive red fluorescent material. For years, the red fluorescing material was also thought to be tremolite, but some or all of it is now thought to be anthophyllite or tirodite. Tremolite from this location also occurs as the variety hexagonite in attractive purple blades. Hexagonite is reported to sometimes fluoresce red under short wave. The activator of the orange or red fluorescence in tremolite is likely to be manganese.

Tremolite is found as flat, gray crystals with a flattened, diamond cross-section in marble at Canaan, Connecticut. These fluoresce lemon yellow under short wave, irregularly and weakly under long wave.

Tremolite crystals of similar appearance are found throughout the Franklin marble of southern New York and adjacent northern New Jersey. Sometimes these are seen as feathery or fibrous gray blades rather than distinct crystals, barely distinguishable on the gray limestone in ordinary light. Under short wave illumination, the fluorescence is pale white with a yellowish, bluish, or greenish cast.

Anthophyllite, $(Mg,Fe)_7Si_8O_{22}(OH)_2$.

Anthophyllite is said to be the predominant red fluorescing amphibole to be found among the fluorescents in the talc-tremolite belt which includes Talcville and Balmat, New York. This mineral is very difficult to distinguish from the other fluorescing amphiboles from this mineral belt, except to note that bladed anthophyllite is said to frequently occur in a rosette pattern. Dense blocks of colorless rosette or radiating crystals fluorescing red or pink are indeed found at the dumps in this area, but dense mats of colorless blades which are grained in one direction (like tremolite) are also found, and these fluoresce in the same red or pink color. The first, and possibly also the second form, is likely to be anthophyllite.

Anthophyllites are generally iron bearing, and those from elsewhere frequently contain up to 20% iron, which is clearly unfavorable for fluorescence. However, the anthophyllites from this area, particularly Talcville, are unusually iron free and contain typically 2 to 3% manganese, while anthophyllite from elsewhere contains little or none. In this, the cause of the red fluorescence may be found.

A fibrous or asbestos-like material, thought to be anthophyllite, is occasionally found at Balmat, fluorescing white. A white fluorescing anthophyllite is reported from Laredo, Texas, and a peach fluorescing anthophyllite from Long Hill, Connecticut, is also reported.

Tirodite, $Mn_2(Mg,Fe)_5Si_8O_{22}(OH)_2$.

Tirodite is another of the glassy amphiboles from the talc-tremolite belt, difficult to distinguish from those previously mentioned. Crystals tend to be large, sometimes slightly pink or slightly tan, often colorless, and they tend to cross in the matrix, jackstraw fashion. However, these tendencies are not dependable guides to identification in all cases. Tirodite is prevalent near Talcville. Manganese seems to be particularly present at this end of the talc belt, and the formation of tirodite is favored since it appears that manganese is an essential constituent of this mineral. Tirodite from this area may contain 10 to 12% manganese, and this may explain the outstanding, intense red or deep pink fluorescence to be seen in these specimens. Large, red fluorescing tirodite crystals may be found in yellow fluorescing talc and may be further accompanied by orange fluorescing tremolite. Such fluorescing specimens are quite outstanding. Seen under long wave light, the best of the tirodites are among the brightest of the red fluorescing minerals.

Pargasite, $NaCa_2(Mg,Fe)_4Al(Si_6Al_2)O_{22}(OH)_2$.

Pargasite from one of the St. Joe mines is reported to fluoresce pale green.

Bustamite, $(Mn,Ca)_3Si_3O_9$.

Bustamite occurs at Franklin, New Jersey, as blocky pink to tan crystals somewhat resembling rhodonite. Under long wave, it sometimes fluoresces a very dull purple-red, but some specimens examined showed a clear and attractive pink fluorescence. The most attractive fluorescing bustamite contains veins or swirls of purple fluorescing hardystonite, providing a nice two-color fluorescent specimen. Some reports also suggest a weak deep red under short wave.

Wollastonite, $CaSiO_3$.

Wollastonite is a product of highly metamorphosed limestones or skarns, and is probably of widespread occurrence. Frequently enough it is a fluorescent, and when it does fluoresce, it is usually in some shade of yellow or orange.

An interesting fluorescent wollastonite from the Harquahala Mountains in Arizona consists of fine, silky, radiating fibers on a nonfluorescent, grainy, white rock. Under short wave, the response is a dull blue-white with some phosphorescence, but under long wave, the response is more varied and more spectacular. Some areas fluoresce yellow and some a bright blue-white, while other areas do not fluoresce at all. The blue-white fluorescing areas produce an attractive and bright green-white phosphorescence which continues to produce light for some time. Wollastonite is found in the Crestmore quarry in Riverside, California, in the form of bundles of white or tan silky fibers. These will fluoresce a mild pink, peach, or yellow, best under long wave ultraviolet.

Wollastonite from Willsboro, New York, containing small green diopsides and pink garnets, may fluoresce dull yellow-white under long wave only.

Wollastonite is found near Hot Springs, Arkansas, as a mass of small, glassy blades. The fluorescence is yellow-orange under long wave, fainter under short wave, with some phosphorescence. The wollastonite is accompanied by fluorescent miserite, which may be an alteration product of the wollastonite. From nearby Magnet Cove, wollastonite occurs which reportedly fluoresces pink under short wave.

Wollastonite from Franklin, New Jersey, is strikingly attractive under short wave. The material may consist typically of areas of wollastonite one-half inch or so in size, fluorescing bright orange under short wave, embedded in calcite which fluoresces orange-red. Sometimes the calcite will also contain small barite crystals that fluoresce blue-gray. This wollastonite has a brief red-orange phosphorescent flash. The response to long wave is tangerine orange. Another combination involves orange fluorescing wollastonite with green fluorescing willemite. Good specimens of either type are rare and exceedingly hard to come by. Many collectors feel that certain of these Franklin wollastonite specimens are the most attractive fluorescents to be seen anywhere in the world. In another mode of occurrence at Franklin, a white fibrous mineral which fluoresces orange, once thought to be pectolite, is now known to be wollastonite.

The Gerstmann collection in Franklin, New Jersey, has what is undoubtedly the largest piece of fluorescing Franklin wollastonite, about 6 inches tall and almost as wide. The material appears to be a section of a large vein several inches wide. Fluorescence is a bright orange. Another wollastonite in that collection shows a habit more typical of non-Franklin specimens: it consists of glassy, gray blades in white calcite, the fluorescence being the usual bright orange. One astonishing specimen contains a round mass of wollastonite which fluoresces fiery orange, surrounded by a partial ring of green fluorescing willemite, surrounded by another ring of violet fluorescing hardystonite, the whole in a mass of red fluorescing calcite – an unbelievable fluorescent specimen!

The Sterling mine at nearby Ogdensburg, New Jersey, has also produced wollastonite in recent years. The wollastonite consists of small blebs up to one-quarter of an inch in size for the most part, densely scattered in calcite. Under short wave in some specimens, the wollastonites fluoresces bright yellow and the calcite does not fluoresce. In others, the calcite forms a pink fluorescent halo around the wollastonite, setting it off very attractively. In yet other specimens, the wollastonite fluoresces orange, while the calcite fluoresces bright pink through the entire specimen, producing an exceptionally attractive combination of warm colors. Some specimens can be found which grade from yellow fluorescing wollastonite at one end to orange at the other with a matching variation in the calcite response. Again, these wollastonites display a phosphorescent flash. In yet other specimens, the band of yellow or orange fluorescing wollastonite blebs merges with a band of green fluorescing willemite blebs of the same size and distribution, providing yet another kind of striking fluorescent wollastonite specimen.

Some of the Ogdensburg material involves larger wollastonite segments which fluoresce orange in a pink to red fluorescing calcite matrix. This can be mistaken for the classic Franklin material, but usually the calcite response is not as red as the Franklin material and not as attractive.

Substantial research has been done on the source of fluorescence in wollastonite. The deep yellow and orange fluorescence is due to the presence of manganese as an activator substituting for calcium atoms in the crystal lattice. However, manganese alone is insufficient to provide a fluorescent response to short wave ultraviolet excitation. Lead must also be present to capture the ultraviolet. The gradation of response from yellow to orange is attributable to a change in concentration of the manganese activator – 2 to 5% producing an orange response, while one-tenth as much produces a yellow response. Interestingly, this same activation system is also at work in the calcite.

Pectolite, $NaCa_2Si_3O_8(OH)$.

The fluorescence of pectolite from Franklin, New Jersey, has been reported to be much the same as wollastonite – orange under short wave light and occasionally paler orange under long wave. A brief phosphorescence has been reported for both. A similar appearance in ordinary light and similar Parker shaft mineral associations have made it difficult to distinguish these minerals. The matter is further complicated by the similarity in fluorescent response of a third mineral, clinohedrite.

Based on recent findings, much of what was thought to be pectolite at Franklin has now been determined to be wollastonite, a closely related pyroxene. Whether there is any fluorescing pectolite at Franklin is now open to some doubt. If it exists, it is scarce. Most material in the Gerstmann collection considered to be pectolite consists of rather formless white masses fluorescing a dull yellow-orange under short wave and dull yellow under long wave ultraviolet, with brief phosphorescence under both. Another specimen is composed of a ball of radiating needles which fluoresce orange, surrounded and enclosed by a perfect ring of red fluorescing calcite, the whole in a nonfluorescent matrix.

Based on its chemical similarity to wollastonite, the orange fluorescence of pectolite is probably due to the presence of manganese.

Pectolite from the trap rocks in the area of Paterson, New Jersey, is sometimes reported to fluoresce yellow. The Baskerville and Kunz report of 1903 suggested that under the iron arc, Paterson pectolite commonly fluoresced in this color. Among a number of specimens examined under modern lights, a few fluoresced dull yellow to strong yellow-orange in areas, best under long wave light. However, close examination shows the fluorescing areas to be altered, possibly to stevensite.

Some pectolite from Paterson fluoresced white to blue-white in areas. As this is found on freshly broken surfaces, there is little reason to doubt that the response is due to the pectolite itself. However, one specimen of pectolite from Secaucus, New Jersey, is distinctly fluorescent. This specimen consists of large, almost glass-clear blades. Under long wave light, the fluorescence is a bright yellow-orange. Under short wave, the response is weak pink. The specimen is associated with small amounts of strong red fluorescing calcite. The fluorescent response is very similar to that of numerous yellow-orange fluorescing wollastonites and scapolites. As in those minerals, manganese may be the activator. The strong red fluorescence of the calcite would support this view.

Yellow or yellow-orange fluorescing pectolite is reported from Magnet Cove, Arkansas, as well as from Lake County and the vicinity of the Golden Gate Bridge in San Francisco, California.

Miserite, $K(Ca,Ce)_4Si_5O_{13}(OH)_3$.

The rare mineral miserite occurs at Wilson Mineral Springs near Magnet Cove, Arkansas. It appears as pink or white blades and masses intimately mixed with wollastonite from which it is difficult to distinguish. Under long wave, what is said to be the wollastonite fluoresces yellow, while the material said to be miserite does not appear to fluoresce. Under short wave, the response of the wollastonite and the miserite is a clear, strong yellow.

Xonotlite, $Ca_6Si_6O_{17}(OH)_2$.

One Franklin specimen of xonotlite examined consisted of diverging, fine, silky white needles. Fluorescence was a weak white under long wave and a weak gray-white under short wave. Xonotlite from Riverside, California, probably the Crestmore quarry, shows a similar fluorescent response.

Margarosanite, $Pb(Ca,Mn)_2Si_3O_9$.

Margarosanite is an elusive and scarce mineral from Franklin, New Jersey, occurring in small quantities probably throughout the ore body. It is elusive in that it appears in different guises, both in ordinary light and as a fluorescent.

In daylight, margarosanite often appears as small, silvery white or pearly, straight, curved, or occasionally swirled blades or plates. Under short wave, it is often seen as spots, veins, or masses fluorescing sky blue of moderate intensity and may be associated with orange fluorescing clinohedrite, orange fluorescing wollastonite, or purple fluorescing hardystonite. Frequently, it is seen disseminated in hyalophane, a barium feldspar. In this form, the silvery or pearly appearance is not evident. The margarosanite is then virtually impossible to recognize by ordinary appearance or hardness, as it takes on the appearance and features of this colorless, dull feldspar. This seems to work a magic transformation in the fluorescence of margarosanite, for such specimens often fluoresce a bright, striking, strong sky blue. One of the most prized of fluorescent specimens is made up of bright blue fluorescing margarosanite in hyalophane, sandwiched between bright green fluorescing willemite, which in turn is sandwiched between bright red fluorescing calcite.

Some margarosanite specimens contain areas which fluoresce red under short wave, at one time thought to be axinite. New data indicate that the red fluorescing areas are also margarosanite. In the Gerstmann collection, there is a magnificent specimen about 5 inches tall, consisting of swirling pearly blades of margarosanite. Under long wave, the margarosanite fluoresces a dark, rich red. Under short wave, some of the blades continue to fluoresce red, while others fluoresce a powdery sky blue. The R. Bostwick collection contains a smaller, straight bladed, but otherwise similar fluorescing margarosanite. Under short wave, the fluorescence is sky blue with some areas fluorescing red. However, under long wave, the blue areas now fluoresce orange rather than red.

Since margarosanite is a lead and calcium silicate in which manganese sometimes substitutes for calcium, one can suspect that the orange and red response is due to manganese.

Walstromite, $BaCa_2Si_3O_9$.

Walstromite, an interesting mineral, comes from the sanbornite deposit in Fresno County, California, which produces so many interesting barium containing minerals. This deposit is found as outcrops for several miles along Rush Creek in the region where Rush Creek merges with Big Creek. Several fluorescent minerals occur here. In addition to walstromite, these include benitoite, fresnoite, and celsian feldspar. Scheelite is reported from nearby, and some weakly fluorescent barite and witherite are reported among the barium minerals.

Walstromite is an analogue of the better known fluorescent, margarosanite. In walstromite, barium takes the place of the lead in margarosanite. The fluorescence of walstromite is pink-orange and rarely pink, under both short wave and long wave ultraviolet. As in the case of margarosanite, manganese substitutes for some calcium in the natural mineral, and it is likely that the fluorescence is due to the presence of manganese. As in wollastonite from the Sterling mine in New Jersey, the pink and orange response may be due to different concentrations of manganese.

Tobermorite, $Ca_5Si_6O_{16}(OH)_2 \cdot 4H_2O$.

A mineral of the Crestmore quarry in Riverside, California, tobermorite is found as white or tan blebs in the Sky Blue limestone. The response under long wave appears to be a yellow-white, but the response is uncertain and more investigation is needed.

Plombierite, $Ca_5H_2Si_6O_{18} \cdot 6H_2O$.

Plombierite occurs as a white, chalky material at the Crestmore quarry in Riverside, California. The response to long wave appears to be cream-white or yellow with phosphorescence, with the short wave response similar but weaker. Plombierite is also reported as fluorescing blue-white under short wave ultraviolet.

Phyllosilicates

Apophyllite, $KCa_4Si_8O_{20}(F,OH) \cdot 8H_2O$.

Apophyllite from the trap rock quarries in and around Patterson, New Jersey, occasionally fluoresces. The fluorescence typically appears in certain spots or areas of the crystal, rather than suffusing the entire crystal. The fluorescent response is sometimes green under either or both short wave and long wave, sometimes yellow-green, and sometimes yellow under long wave. A yellow response has been reported under long wave in apophyllite from the Crestmore quarry in Riverside, California.

Pyrophyllite, $Al_2Si_4O_{10}(OH)_2$.

Pyrophyllite from Staley, North Carolina, and from Graves Mountain in Georgia fluoresces butter yellow under long wave ultraviolet.

Talc, $Mg_3Si_4O_{10}(OH)_2$.

Talc is found, commonly mixed with tremolite, in a line of mines stretching from Balmat to Talcville in St. Lawrence County, New York. The fluorescence of this talc is best under short wave ultraviolet and appears as bright lemon yellow or as yellow with a slight green cast. Sometimes a faint fleeting phosphorescence is evident.

At Balmat, in the open pit mine, narrow parallel bands of fluorescing talc are seen. In places, as the short wave light scans upward across these bands, the color of the fluorescence gives way to orange, indicating the parallel bedding of fluorescent tremolite. At Talcville, blocks of orange fluorescing tremolite are found bearing layers of yellow fluorescing talc. Here also, talc is found as the matrix for red fluorescing tirodite crystals. Sometimes, fluorescent talc, tremolite, and tirodite are found together to provide outstanding specimens.

Talc is also found in the mine at Franklin, New Jersey, is an unusual form. Rather than its usual flaky form, this material is compact, smooth, and almost glassy in appearance. The fluorescence is yellow under long wave ultraviolet.

Gyrolite, $Ca_2Si_3O_7(OH)_2 \cdot H_2O$.

Gyrolite is found as satiny, white, radiating fibers in the Crestmore quarry, Riverside, California. The fluorescence is green or occasionally blue-green, under both long and short wave. In view of the frequent occurrence of hyalite in this quarry, the green fluorescence of gyrolite may be due to a surface alteration of gyrolite to hyalite. Gyrolite from New Almaden, California, is reported to fluoresce brown under short wave.

Phlogopite, $K(Mg,Fe)_3(AlSi_3)O_{10}(F,OH)_2$.

Phlogopite is a magnesium biotite commonly found in metamorphosed limestones. At Sparta and Franklin, New Jersey, it appears in bands in the Franklin marble in a bronze or orange color, but may be colorless. The fluorescence is usually straw to lemon yellow under short wave, particularly evident at edges and fractures in the mica flake. It is sometimes accompanied by a parallel band of norbergite which fluoresces a somewhat stronger orange-yellow. It is also occasionally found with fluorescent edenite, forming a nice two-color fluorescent.

Phlogopite also occurs in the limestones of northern New York State. The material from Newcomb, New York, is found in the limestone as small, colorless, silvery flakes which accompany small, colorless grains of diopside. Under the short wave lamp, the phlogopite fluoresces bright lemon yellow, while the diopside fluoresces bright sky blue. The combination furnishes an extremely attractive specimen.

Margarite, $CaAl_2(Al_2Si_2)O_{10}(OH)_2$.

Margarite is found as whitish or blue-white blades resembling mica within the calcite or Franklin marble at the Sterling mine, Ogdensburg, New Jersey. The fluorescence is a moderate to strong sky blue, best under long wave. Some margarite is found with corundum, providing attractive two-color fluorescent specimens.

Deweylite.

Really an intimate mixture of serpentine and disordered talc rather than a true mineral, deweylite is a yellowish tan material found in veins and fissures in serpentines. Deweylite from the many quarries along the Maryland-Pennsylvania border fluoresces light yellow, best under long wave ultraviolet, and is occasionally phosphorescent. A similar response is seen in deweylite from the Crestmore quarry in Riverside, California.

Hectorite, $Na_{0.33}(Mg,Li)_3Si_4O_{10}(F,OH)_2$.

A lithium clay of bright white color, hectorite is found at Hector, California. Long wave fluorescence is bright sky blue, and the short wave response is similar but weaker. Tan crystals which fluoresce bright yellow, probably colemanite, are often found in this material.

Prehnite, $Ca_2Al_2Si_3O_{10}(OH)_2$.

Numerous specimens of prehnite, a green mineral from the Paterson, New Jersey area trap rocks, show a faint, indistinct purple-white color under long wave. This is a doubtful but possible fluorescence.

Prehnite from Franklin, New Jersey, is found as closely packed, glistening white crystals accompanied by a number of other minerals typical of the famous Parker shaft. Under short wave, the prehnite fluoresces a mild peach, giving way to purple in some areas, possibly due to visible light leaking through the filter of the ultraviolet light. Clear tan crystals said to be prehnite are found in the pegmatite at Crestmore quarry, Riverside, California; these fluoresce blue-white under short wave and yellow under long wave.

Searlesite, $NaBSi_2O_5(OH)_2$.

Some searlesite from the Green River district, Sweetwater County, Wyoming, fluoresces orange under long wave ultraviolet.

Tectosilicates

Quartz, SiO_2.

A number of distinctive forms are grouped under the name quartz, having the same chemical formula but differing in network structure.

These include quartz in the well-known crystalline form, cryptocrystalline materials such as chalcedony, and glass-like types such as opal.

In spite of its widespread occurrence, there is little evidence for the fluorescence of crystalline quartz. Quartz crystals in geodes often appear to fluoresce the green typical of uranium activation, but in many cases, close examination indicates that it is the underlying layer of chalcedony which is responding to the ultraviolet light. In the case of geodes from Iowa which are lined with sugary druses of quartz, the crystals appear to fluoresce. However, the microscope reveals as the source of fluorescence a center spindle in the crystal of what appears to be a noncrystalline form of quartz. On the other hand, masses of a milky quartz from the feldspar pegmatites of Spruce Pine, North Carolina, may fluoresce a weak straw yellow under long wave. Such quartz is probably crystalline in internal structure, though not in external form. Also, reports exist in the literature of amethyst from North Carolina fluorescing blue. Certainly, a better confirmation of fluorescence in crystalline quartz would be desirable.

Chalcedony, commonly found as a lining in geodes, usually fluoresces green due to the presence of uranium. Also, butter yellow, and even a surprising orange or blue, have been found in chalcedony under long wave. Common opal from the western and mountain states also fluoresces green. Large pieces of a white common opal from Virgin Valley, Nevada, are particularly outstanding in this regard. All such green fluorescences are attributed to uranium in concentrations as low as one-hundredth of one percent. In a night prospect on some deserts of the West or the Badlands of South Dakota, the ground lights up under the ultraviolet light in innumerable spots of bright green due to one or the other of these materials.

Hyalite, like opal, is a noncrystalline form of quartz. It occurs in crevasses in pegmatites as a layer or mass of small, colorless, clear, glassy beads, delicately bonded together. Hyalite from Hyalite Mountain, Montana, fluoresces green due to uranium, as does hyalite from near Middletown, Connecticut, and from other pegmatite quarries throughout New England. Very brightly fluorescing hyalite is found frequently throughout the Crestmore quarry in Riverside, California. In its fluorescent brilliance, this hyalite far outshines the more subdued fluorescences of the other minerals of Crestmore. The claim to the most brilliant hyalite fluorescence is probably reserved to the material from Spruce Pine, North Carolina. This may be the brightest of all fluorescent material, fluorescing brilliant green and rivaling the best of the willemite from Franklin, New Jersey, in luminous intensity.

Additional information on quartz will be found in Part B of this chapter.

Tridymite and Cristobalite, SiO_2.

These high temperature forms of quartz are found as small crystals in cavities in hardened lavas. Tridymite is found in obsidian from Obsidian Cliff, Yellowstone Park, Wyoming, fluorescing dark cherry red under short wave. Cristobalite has been reported from various locations as fluorescing either green or red. Further investigation would seem warranted to confirm the identification and fluorescence of such materials.

Melanophlogite, SiO_2.

Melanophlogite is found at Mount Hamilton, Santa Clara County, California, in clear, glassy crystals of a cubic figure. Under long wave ultraviolet, the response is a weak gray-white, brightening to yellow in areas which appear to contain an admixture with some other material. The short wave response is similar but weaker.

Feldspar $(K,Na,Ba)Al(Si,Al)_3O_8$.

The feldspars are the most common and widespread group of rock forming minerals. Yet, despite their wide occurrence, it is not widely recognized that they are often fluorescent. The fluorescence seems to be found particularly in feldspars from pegmatites, and most such fluorescent specimens appear to be microcline.

In quarries around Middletown, Connecticut, some feldspar fluoresces violet-blue or purple under short wave, other specimens fluoresce sky blue under long wave, and some specimens contain both blue and purple areas. Such feldspar is sometimes found with yellow fluorescing apatite and green fluorescing hyalite, providing very interesting two-and three-color fluorescent combinations. Fluorescent microcline is found with garnet as pods in schist at Lime Crest quarry, Sparta, New Jersey. The fluorescence is a vivid violet-blue under short wave. Under long wave, some of these areas remain violet-blue, while others fluoresce sky blue. A similar microcline is found at Franklin, North Carolina, in schist with rhodolite garnet, fluorescing deep violet blue under short wave and long wave. All of the above responses are so similar as to suggest an identical cause, perhaps a rare earth activator, most probably europium.

In Sparta, New Jersey, a microcline is found which fluoresces weak dark red under short wave. An apparently identical fluorescent feldspar is found on the dumps at nearby Franklin, New Jersey. A brown feldspar found at Boron, California, fluoresces an identical dark red. Iron has been implicated as an activator of this red fluorescence.

Both violet-blue and red fluorescing microcline feldspars are found in the pegmatite at the Crestmore quarry in Riverside, California.

Green amazonite microcline found at Franklin, New Jersey, fluoresced a weak gray-white or blue-gray under short wave, while a Franklin specimen in the Paterson Museum labeled oligoclase fluoresced a good yellow under long wave only. Franklin, as would be expected, has more exotic feldspars, including hyalophane reported to fluoresce weak red under short wave and also a bright blue fluorescing hyalophane whose response is said to be due to included margarosanite. Celsian feldspar from Franklin fluoresces light violet-blue under short wave.

A similar light violet-blue fluorescing celsian is also found at the sanbornite deposit in Fresno County, California, the source of fluorescent walstromite.

Albite from Liepers quarry in Swarthmore, Pennsylvania, fluoresces yellow-white, best under long wave; feldspar, possibly albite, from Kings Mountain, North Carolina, containing spodumene crystals, fluoresces yellow or blue-white.

World famous amazonite microcline crystals are found in the Pikes Peak area of Colorado. In ordinary light, these may be green at the top and brown at the roots. Fluorescence under long wave is green at the top, but bright yellow in the root areas. A similar green and brown amazonite from Media, Pennsylvania, has a similar green and yellow fluorescence. A microcline from Custer, South Dakota, fluoresces an attractive bright sky blue under short wave. An albite feldspar from Magnet Cove, Arkansas, is reported to fluoresce weak pink under short wave. Andesine feldspar from Washington Pass, Washington, fluoresces pale white under long wave ultraviolet and is sometimes found with the rare fluorescent zektzerite.

Mention has been made of feldspar which may fluoresce violet-blue under short wave and sky blue under long wave, but some feldspars show a more dramatic two-color fluorescent response. Feldspar fluorescing red under short wave is found intermixed with magnetite at the Edison mine in Sussex County, New Jersey. Under long wave, some of the parts which fluoresced red now fluoresce blue. These feldspars are found with pink-orange fluorescing apatite and with fluorite which fluoresces blue under long wave and yellow under short wave. The blue fluorescence of both the feldspar and the fluorite and the yellow fluorescence of the fluorite are undoubtedly due to rare earths which are plentiful in these magnetite deposits.

In view of the wide variety of fluorescent responses seen in feldspar, further study would be desirable in order to better identify the feldspar species involved and the activators responsible.

Additional information on feldspar will be found in Part B of this chapter.

Sodalite and Hackmanite, $Na_4Al_3(SiO_4)_3Cl$.

Sodalite is probably more familiar to the lapidary specialist than to the specimen mineral collector, as it often has an attractive blue color reminiscent of lapis lazuli. It is also found in other colors, particularly white and gray. Hackmanite is a variety of sodalite exhibiting the astonishing property known as tenebrescence or photochromism. A freshly broken piece of hackmanite will show a pink-purple to red surface. After a few minutes of exposure to visible light, this color will more or less completely fade away, revealing the white surface ordinarily characteristic of the mineral. On exposure to a few minutes of short wave ultraviolet, the color will return as strong as before. This cycle of color change can be repeated indefinitely.

Outstanding sodalite or hackmanite specimens are not found in this country which can compare with well-known Canadian material. Fluorescent sodalite is an abundant constituent of the nepheline-syenite dike at Beemerville, New Jersey, just off the Appalachian Trail. The sodalite consists of small irregular white or gray flecks in a crumbly syenite host rock. The fluorescence is orange under long wave, and the sodalite is accompanied by small crystals of yellow fluorescing zircon. Blue sodalite is found at Magnet Cove, Arkansas, and is reported to fluoresce violet-red; the ultraviolet band is not stated. Hackmanite is also found there and is reported to fluoresce yellow under long wave and rose under short wave. As the hackmanite designation indicates, this material is tenebrescent. It is not reported whether it behaves like Canadian hackmanite, in which the fluorescent portions of the specimen are usually also the tenebrescent areas and which may also display a faint silver-blue phosphorescence in the same areas.

Orange fluorescing sodalite also occurs at Red Hill, Moultonboro, New Hampshire. This fluoresces best under long wave but may fluoresce weakly under short wave. Work with synthetic sodalite indicates that it is the presence of sulfur as a polysulfide which is responsible for the fluorescence of sodalite.

Scapolite and Wernerite, $(Na,Ca)_4Al_3(Al,Si)_3Si_6O_{24}(Cl,CO_3,SO_4)$.

Scapolite is the name of a series of minerals in which the members differ from one another usually by a smooth and continuous substitution of constituent elements. Wernerite is the informal name of the mineralogical middle ground of this set, and sometimes the name scapolite is used in the same sense. The usual mode of occurrence is as gray or white masses or crystals locked in the host rock, usually a metamorphosed impure limestone or skarn. The fluorescent response is variable, though there is no evidence available that this is due to any particular variation of scapolite composition.

The most famous fluorescent scapolite is the yellow fluorescing wernerite from Canada, and the United States has little to match it in consistency or brilliance of response. However, at one time a deep blue material which proved to be lapis lazuli was recovered from the St. Joe zinc mine at Edwards, New York, in a matrix of colorless material. This colorless material has proven to be the dipyre variety of scapolite. Under long wave in particular, this fluoresces orange-yellow, matching in intensity and color the best of the Canadian wernerites.

The scapolites found in the Franklin marble at Sparta, New Jersey, are probably more typical. Under long wave ultraviolet, some specimens fluoresce pale yellow, some orange, and others pink. Under short wave, all of these responses revert to a washed out yellow-white. One specimen, however, showing no response under long wave, produced a beautiful violet-pink response under short wave. All of these phosphoresce weakly under stimulation by both lights. A generally similar variety of fluorescent responses is found in scapolite from the Crestmore quarry in Riverside, California.

From the mine at Sterling Hill, New Jersey, there is reportedly a scapolite in calcite fluorescing dull yellow under both long and short wave with a good phosphorescence. Scapolite from the pegmatite contact at Franklin, New Jersey, may be associated with allanite, microcline, pyroxene, and others, and may fluoresce peach-pink and occasionally deep sky blue under long wave. A short wave blue fluorescing scapolite is reported in red fluorescing calcite and dolomite at Franklin.

Scapolite is also reported to have been found in the magnetite mine at French Creek, Pennsylvania, fluorescing dull white under long wave. It is reported also from Danbury, Connecticut, fluorescing yellow to pink, with associated minerals similar to that of the pink fluorescing scapolite from Franklin. Nice free-standing crystals of scapolite from Pierrepont, New York, fluoresce pink to dark red under short wave. Finally, probably the oldest report on this subject derives from the Kunz and Baskerville report of 1903, in which they described phosphorescent wernerite from New York; the color and precise location were not given.

Research indicates that, as with sodalite, paired sulfur atoms constitute the activator in the fluorescent yellow scapolites (wernerite) from Canada. This may be the activator of the yellow fluorescent scapolites described here, and may also play a role in the orange and pink responses.

Additional information on scapolite will be found in Part B of this chapter.

Analcime (Analcite), $NaAlSi_2O_6 \cdot H_2O$.

Analcime appears to be a reluctant and very occasional fluorescent. Crystals of analcime from the trap rock quarries of Paterson, New Jersey, sometimes fluoresce a mottled yellow under long wave and, with less certainty, a weak blue-white under short wave. From Golden, Colorado, analcime sometimes fluoresces green in some areas under both long and short wave, and it is also reported to fluoresce cream-white under long wave.

Pollucite, $(Cs,Na)_2(Al_2Si_4)O_{12} \cdot H_2O$.

Pollucite, a cesium bearing zeolite, is occasionally reported as a fluorescent, with reports differing on what the fluorescent response is. Pollucite from pegmatite near Branchville, Connecticut, is reported to fluoresce an attractive blue-green under short wave only, while material from Newry, Maine, is said to fluoresce weak orange under long wave, and material from Custer, South Dakota, cream-white under short wave.

Natrolite (and Mesolite), $Na_2Al_2Si_3O_{10} \cdot 2H_2O$.

Natrolite from the trap rocks in the Paterson, New Jersey, area sometimes fluoresces, though unevenly. The response seems to be concentrated in the root areas of these needle-like crystal groups. The fluorescent color can be yellow-white or blue-white, with the short wave response being stronger in some specimens while the long wave is stronger in other specimens. More often, specimens from this area do not fluoresce at all. It is possible that the specimens examined are really mesolite, since a number of mesolites from Paterson, particularly the needle-like specimens, have been mislabeled as natrolite.

Benitoite crystals from San Benito County, California, are found in a bed of natrolite in serpentine rock. The natrolite is often partially etched away in order to reveal the benitoite in the best way. This natrolite fluoresces white under long wave, but it is possible that this response is due to an alteration resulting from chemical etching.

Additional information on natrolite will be found in Part B of this chapter.

Laumontite (and Leonhardite), $CaAl_2Si_4O_{12} \cdot 4H_2O$.

Laumontite from Leesburg, Virginia, fluoresces a weak white under both long and short wave lights. A similar fluorescent laumontite has been found in the magnetite mine at Morgantown, Pennsylvania, and in various zeolite prospects of Paterson, New Jersey.

Laumontite fluoresces more frequently than other zeolites, and perhaps half the specimens seen may fluoresce to one degree or another. It is possible that much or all of this material is leonhardite, or a leonhardite coating, since leonhardite is a frequent dehydration product of laumontite.

Stilbite, $NaCa_2(Al_5Si_{13})O_{36} \cdot 14H_2O$.

Stilbite is one of those very reluctant and infrequent fluorescents so typical of the zeolites. White stilbite from Thomaston, Connecticut, fluoresces weak white under long wave ultraviolet. The response is uneven over the specimen and appears to be confined to a thin surface layer, which suggests that the response is due to some form of surface alteration.

Heulandite, $(Na,Ca)_2Al_3(Al,Si)_2Si_{13}O_{36} \cdot 12H_2O$.

Occasionally, heulandite is found as a fluorescent, and material from the Paterson area sometimes fluoresces yellow-white under long wave. Heulandite from Jefferson County, Colorado, is also reported to sometimes fluoresce white under long wave.

Chabazite, $CaAl_2Si_4O_{12} \cdot 6H_2O$.

Like other zeolites, chabazite is only an infrequent and occasional fluorescent. Some specimens of these somewhat squarish crystals exhibit an interesting and unexpected response. Thus, chabazite crystals from Paterson, New Jersey, sometimes fluoresce a bright green under both short and long wave, but brightest under short wave. This response appears throughout the entire crystal and provides an attractive fluorescent specimen.

Cowlesite, $CaAl_2Si_3O_{10} \cdot 5\text{-}6H_2O$.

Cowlesite is found as white, crystalline blades filling cavities in lava at Beach Creek, Grant County, Oregon. The fluorescence is a not very intense white or gray-white under both long and short wave.

Thomsonite, $NaCa_2(Al_5Si_5)O_{20} \cdot 6H_2O$.

Thomsonite from the trap rocks in the area of Paterson, New Jersey, is an occasional and reluctant fluorescent. The response is a weak and uneven white under long wave.

At Franklin, New Jersey, thomsonite occurs as a dull, chalky white material of no particular attractiveness. Under either short wave or long wave light, the response is about the same. The fluorescence is a mild white, while what appears to be a surface alteration or coating responds a mild blue-white. It is also possible that these specimens are xonotlite rather than thomsonite.

Esperite, $(Ca,Pb)ZnSiO_4$.

Once called calcium-larsenite, esperite is one of the numerous fluorescent silicates containing calcium, zinc, or lead in some combination, taken out through the Parker shaft at Franklin, New Jersey. Esperite is found as masses, and crystals may not exist. These masses are sometimes white or a dirty white, but more often esperite has a slight tan-pink appearance. It is frequently accompanied by nonfluorescent masses of franklinite, and at the interface between these two, a diffuse band of red material is usually seen. This may be red lead oxide, but it is more likely the red zinc oxide, zincite. From this band, the red mineral diffuses or marbles through the esperite, and this may be the origin of the frequent tan-pink color.

Esperite fluoresces a brilliant, intense lemon yellow under short wave ultraviolet. Response under long wave is a not particularly attractive dull yellow to yellow-green. A frequently associated fluorescent is hardystonite, and an even more attractive one is fluorescent willemite. Willemite and esperite fluoresce with equal intensity, and the two fluorescent colors, yellow esperite and green willemite, make a startling and unforgettable combination. Esperite and hardystonite or willemite occur as interlocked blocky segments, but sometimes they are combined in alternating bands or in strange fluorescent whirls and swirls, as if stirred together, but only incompletely. Sometimes also, esperite can be seen in yellow fluorescent bands in red fluorescing calcite, and sometimes willemite is again also present. The ultimate fluorescent is seen when one specimen contains esperite, willemite, hardystonite, clinohedrite, and calcite, fluorescing an intense yellow, green, blue, orange, and red under short wave.

The cause of the fluorescence in esperite – the particular activator or activators – is not known. Esperite contains lead as a main constituent metal, and since a large variety of different lead minerals fluoresce yellow at least occasionally, one can suspect that lead plays some role in the fluorescence of esperite. However, based on the similarity between the fluorescent spectra of esperite and willemite [see Figure 9-8(a)], the same activator – manganese – can equally well be suspected.

Esperite is a scarce mineral and highly prized in every fluorescent collection which contains it; it is particularly prized in the fluorescent combinations mentioned above.

Other Silicates

Allophane.

A bluish material, allophane is a frequent fluorescent. Allophane from the Gold Hill mine in Toole County, Utah, fluoresces blue or blue-green under short wave, better under long wave. Excellent green fluorescing aragonite is also found at this mine. Allophane from Resin Hollow, Kentucky, is reported to fluoresce white or green.

Additional information on allophane will be found in Part B of this chapter.

Wickenburgite, $Pb_3CaAl_2Si_{10}O_{24}(OH)_6$.

A lead and calcium silicate, wickenburgite is found in several locations in Arizona, the original site being the Potter-Cramer mine. Material from the Rat Tail Claims appears as colorless, transparent, microscopic, tabular crystals, in a mixture of numerous other minerals. The fluorescence is dark red or pink orange under short wave, with apparently no fluorescence under long wave. It is found with an unidentified mineral which fluoresces cornflower blue under short wave only, and with a yellow-green fluorescent which is probably willemite. Under long wave, ample amounts of blue fluorescing fluorite are visible, and under both bands, red fluorescing calcite can be confused with the similar fluorescence of the wickenburgite. Numerous other nonfluorescent but colorful minerals are also in evidence. The combination of fluorescents makes the typical specimen quite attractive.

Brannockite, $KSn_2Li_3Si_{12}O_{30}$.

Brannockite, a silicate of lithium and tin, is a rare mineral occurring in small amounts in a tin and lithium rich pegmatite at Kings Mountain, North Carolina. It is reported to fluoresce bright blue-white under short wave.

Fresnoite, $Ba_2TiSi_2O_8$.

A rare mineral, related to benitoite in the sense that it is composed of the same elements as benitoite but in a different atomic ratio and different molecular structure, fresnoite is found with benitoite traces at the Victoria mine, San Benito County, and in the Rush Creek area of Fresno County, California. The fluorescence is bright yellow to yellow-green under short wave. Under ordinary light, this mineral appears as blotches of tan, granular material in the host rock.

Zektzerite, $NaLiZrSi_6O_{15}$.

Zektzerite is a rare mineral from Washington Pass, Washington. It occurs as well-formed, stubby, white crystals – typically up to an inch in size. Fluorescence is a positive and strong white under short wave.

Rosenhahnite, $3CaSiO_3 \cdot H_2O$.

Found as a thin, white, crystalline coating over the surface of a dark gray host rock along Highway 101 in Mendocino County, California, rosenhahnite has a long wave fluorescence that appears to be white with a decided yellow-green tint. The short wave response is less well defined, but whitish. However, since rosenhahnite from this location is often mixed with any of seven other minerals which also happen to be potentially fluorescent, confirmation that the fluorescence is due to rosenhahnite is desirable.

Rosenhahnite is also reported from the Durham quarry in Wake County, North Carolina with a light orange-pink response under long wave ultraviolet and no response under short wave.

Jonesite, $Ba_4(K,Na)_2Ti_4Al_2Si_{10}O_{36} \cdot 6H_2O$.

Found as small, almost microscopic, crystals in the benitoite mine in San Benito County, California, jonesite is reported to fluoresce dull orange under short wave, with apparently no response under long wave.

HYDROCARBONS

Karpatite, $C_{24}H_{12}$.

Also spelled carpathite and referred to as pendletonite, karpatite has been found to be identical to an organic material known as coronene.

Karpatite is a rare hydrocarbon found in the Fourth of July mercury mine near Idria, San Benito County, California. Organic substances are not the favorites among the authors of mineral books, who may favor rockier substances. Yet karpatite is of natural occurrence, has a well-defined composition, is crystalline, and should therefore be considered a mineral. It is found as small, yellow-green crystals, a few sixteenths of an inch in size, diverging along the surface of the host rock which appears to be an iron stained sandstone. Red flakes of cinnabar are a frequently associated mineral. The fluorescence is a vivid electric blue, or sometimes blue-green, under both short and long wave ultraviolet light.

Idrialite and Curtisite, $C_{22}H_{14}$.

Like karpatite, idrialite and curtisite are fluorescent organic minerals. For a time it was thought that idrialite and curtisite were identical, but current research indicates otherwise. Idrialite from Skaggs Springs, California, golden yellow under visible light, fluoresces a brilliant orange-yellow, green, or yellow-green under both long and short wave ultraviolet. Curtisite is also found at Skaggs Springs and is reported to fluoresce green.

Idrialite is also reported from the karpatite locality at Idria, California, with a yellow-green fluorescence.

Other Hydrocarbons

Hydrocarbons in milky quartz from Ilwaco and from Snohomish, Washington, fluoresce in blues, yellows, and white. Such hydrocarbons may include idrialite or curtisite, or may be of a different composition. Hydrocarbons appear to activate yellow-white fluorescing fluorites, including those from Rosiclare, Illinois, and Clay Center, Ohio, and may activate other minerals as well.

URANIUM MINERALS

This section treats a group of fluorescent minerals, most of which fluoresce in much the same green color and which share a common activator, uranium. In these minerals, uranium is not a trace element or an impurity but plays an essential role in defining the mineral species. In contrast, uranium is present in trace amounts in other minerals where it is not an essential constituent but an impurity. In such minerals, hyalite opal for example, it causes a similar green fluorescence. Such fluorescences are described under the appropriate host mineral name elsewhere.

Uranium is incorporated as a uranyl group, a combination of one uranium atom with two oxygen atoms (UO_2) in a tightly bound, stable combination. Typically, the minerals treated here are composed of a metal, a uranyl group, and either a carbonate, phosphate, arsenate, sulfate, or silicate group. Also, a hydroxyl group (OH) may be present. It should be noted that in addition to the uranyl group, a water "molecule" is almost always present in these fluorescent minerals and may also play some role in fluorescence. Some hydrated oxides containing uranyl are also included in this section.

There have been attempts to establish rules for the degree of fluorescence of these minerals. These rules are usually found to be valid, but not always so. Thus, the rule states that calcium or barium uranyl carbonates, phosphates, or arsenates are strongly fluorescent, while magnesium equivalents are less strongly fluorescent. Copper, iron, lead, or bismuth equivalents are only weakly fluorescent or nonfluorescent. Potassium or calcium uranyl silicates also fluoresce weakly or are nonfluorescent. Many apparent contradictions to these rules may result from impurities, for it is probably impossible to find one of these minerals entirely free from another. Thus, torbenite, a copper uranyl phosphate, should not fluoresce, but specimens are found which do fluoresce. This is attributed to the presence of some autunite, a calcium uranyl phosphate, which is a fluorescent. Also, different degrees of hydration may affect fluorescent brightness since these minerals are zeolite-like and able to increase or decrease water content.

There have also been attempts to establish rules for fluorescent color. These rules have not been very useful. The fluorescent color of these minerals is very difficult to describe satisfactorily in any case, and the slight differences in their shades of fluorescent color even more so, and different authorities have disagreed in their choice of descriptive words. Also, like intensity, color may shift somewhat due to mixing of minerals usually found in nature and due to different degrees of hydration. It is common then to refer to the fluorescent color of most of these minerals as "yellow-green." All collectors know the yellow-green color since it is the characteristic "uranium" fluorescent color, seen most commonly in fluorescent hyalite and the like. This color is so characteristic that when it is seen under ultraviolet light, it can usually be stated immediately that the color is due to the presence of uranium, either as an essential element of the mineral or as a trace impurity. Among widespread minerals, only willemite is similar in fluorescent color and might be confused with uranium.

However, not all of the large list of fluorescing uranium minerals respond in quite this way. In particular, the uranium bearing carbonates either fluoresce in a different shade or are variable in fluorescent color, since various authorities describe their fluorescence differently. Thus, andersonite, liebigite, swartzite, and the carbonate-sulfate schroeckingerite are at different times all described as fluorescing yellow-green, white-green, blue-green, or just green.

Specimens of these minerals examined by the writer do fluoresce differently from the yellow-green characteristic of most uranium bearing minerals. For example, when andersonite is directly compared with autunite under ultraviolet, the autunite fluoresces intense yellow-green while the andersonite fluoresces in a shade of green devoid of yellow. This shade is generally the color of a green traffic light. The direct comparison makes the andersonite appear bluish due to the contrast with the yellow in autunite, but when examined alone, the andersonite does not appear particularly blue. Andersonite specimens do appear to vary in the degree of whiteness of the fluorescent response, some being somewhat more green-white rather than pure green.

Spectrum analysis of the fluorescent light produced by various uranium minerals provides some explanation of the differences seen, and this is discussed in chapter 9.

These fluorescent greens usually identify a mineral as uranium bearing, but ultraviolet fluorescence serves poorly as a means to distinguish between these minerals, the colors either being substantially identical from mineral to mineral or differing in ways which cannot be relied upon as a means of specific identification. Because of the similarity of fluorescent response and the similarity of daylight color and appearance, these minerals are probably subject to more than the ordinary degree of misidentification and mislabeling in collections.

Interestingly, the uranium minerals of the greatest commercial importance, coffinite, uraninite, and carnotite, do not fluoresce. However, they are frequently surface altered to one of the fluorescing uranium minerals detectable under ultraviolet light. As a field test, a sample can be sprayed with an acid which brings about a quick alteration, acetic acid being particularly effective for this purpose.

Finally, there are a few uranium minerals whose fluorescence is not of the green uranyl type. This fluorescence may be due to a uranyl activator whose energy levels are distorted by the electric fields of nearby atoms or to another cause entirely.

Uranyl Oxides and Hydroxides

Schoepite, $UO_3 \cdot 2H_2O$.

Reported from Beryl Mountain in New Hampshire and from various mines in the San Rafael Swell, Utah, schoepite reportedly fluoresces bright yellow-green.

Becquerelite, $CaU_6O_{19} \cdot 11H_2O$.

Becquerelite has been reported from the San Rafael Swell, the White Canyon district, and elsewhere in Utah; from Monument Valley in Arizona; and from the Jackpile mine near Grants, New Mexico. Its fluorescent response is reported as dull brown or more often nonexistent.

Uranyl Carbonates

Andersonite, $Na_2Ca(UO_2)(CO_3)_3 \cdot 6H_2O$.

Andersonite is found as thick tufts and effervescences on matrix in the Ambrosia Lake district near Grants, New Mexico. The color in ordinary light is white with a distinct lime green tint. The fluorescence is a brilliant green, green-white, or green-white with a hint of blue, under either short or long wave. Andersonite is also reported from the Skinny No. 1 mine, Thompsons district, Utah, and from the Hillside mine in Yavapai County, Arizona.

Liebigite, $Ca_2(UO_2)(CO_3)_3 \cdot 10H_2O$.

Liebigite has been reported from the Pumpkin Buttes area, the Lucky Mt. mine, and Lusk, Wyoming; and from the Black Ape mine, Thompsons district, Utah. Its reported fluorescence is much the same as andersonite. It is reported from Jim Thorpe, Pennsylvania, as a blue-green fluorescent.

Rutherfordite, $UO_2(CO_3)$.

Reported from Beryl Mountain, New Hampshire, and Newry, Maine, as an alteration of uranite, rutherfordite is also found with gummite at the Ruggles and Palermo mines in New Hampshire. The fluorescence is reported as a weak yellow-green or none.

Swartzite, $CaMg(UO_2)(CO_3)_3 \cdot 12H_2O$.

Swartzite, reported from the Hillside, Yavapai County, Arizona, with other uranium minerals, is reported to have a bright yellow-green fluorescence.

Rabbittite, $Ca_3Mg_3(UO_2)_2(CO_3)_6(OH)_4 \cdot 18H_2O$.

Rabbittite may have been a one-time occurrence on pitchblende from San Rafael Swell, Emery County, Utah. Its fluorescence is reported as weak.

Bayleyite, $Mg_2(UO_2)(CO_3)_3 \cdot 18H_2O$.

Reported from the Hillside mine in Arizona, with other uranium minerals, from the Hideout mine in the White Canyon district in Utah, from some places in Wyoming including streambeds in the Pumpkin Buttes area, from Colorado at the Rifle and P.J. mines, and from the Grants region of New Mexico, bayleyite is reported to fluoresce a weak yellow-green.

Schroeckingerite, $NaCa_3(UO_2)(CO_3)_3(SO_4)F \cdot 10H_2O$.

Reported from the Hillside mine in Arizona, from several mines in the Thompsons district, from Marysvale, from the Hideout mine in the White Canyon district, Utah, from the Homestake claims near Grants, New Mexico, and from numerous other locations in the West, schroeckingerite has also been reported from Jim Thorpe, Pennsylvania. The fluorescence is usually reported as yellow-green, sometimes as bluish or whitish green. It is usually associated with the other minerals of this list. At one time it was called dakeite after a man who contributed much to the study of fluorescent minerals.

Uranyl Sulfates

Zippeite, $K_4(UO_2)_6(SO_4)_3(OH)_{10} \cdot 4H_2O$.

Zippeite from the Ambrosia Lake district in New Mexico occurs as little balls of bright yellow color scattered sparsely over a rock matrix, or as a smooth coating like a canary yellow paint. The fluorescence is a bright green or yellow-green. It occurs at numerous other localities including mines of the San Rafael Swell, Marysvale, and White Canyon areas in Utah, and various locations on the Colorado Plateau. It is reported that some zippeites do not fluoresce. Zippeite can occur in numerous states of hydration, and it is possible that this is responsible for the variation in fluorescent response.

Uranopilite, $(UO_2)_6(SO_4)(OH)_{10} \cdot 12H_2O$.

In uranopilite reported from the White Canyon and Marysvale, districts in Utah, the fluorescence is reported as bright yellow-green which disappears with dehydration of the mineral to a meta form. This mineral is also reported from the Laguna district in New Mexico.

Uranyl Phosphates and Arsenates

Autunite and Meta-autunite, $Ca(UO_2)_2(PO_4)_2 \cdot 2\text{–}6H_2O$.

The finest specimens of autunite have come from Mt. Spokane, Washington, as yellow-green or dark grass green mica-like crystals. The dominant yellow color of some specimens suggests the meta phase, that is, one in which some dehydration has taken place. This is the phase usually found in the natural mineral. The fluorescence for either phase is a brilliant yellow-green under short or long wave.

Less spectacular specimens are found at Marysvale, Utah, and throughout the pegmatites of North Carolina and New England. Smallish yellow crystals are found on white feldspar in the Spruce Pine, Little Switzerland, Burnsville area of North Carolina, fluorescing bright yellow-green and sometimes mixed with hyalite. Similar occurrences are reported from the Ruggles mine in Grafton Center, New Hampshire, and from other localities, particularly in the West, too numerous to mention.

It is not always possible to know what specimens are autunite and which meta-autunite, so that distinctions in fluorescence, if any, are not easily recognized.

Uranospinite, $Ca(UO_2)_2(AsO_4)_2 \cdot 10H_2O$.

Reported from Kane County in Utah, with autunite in the Gas Hills region of Wyoming, and with torbernite from the Golden Gate Canyon area of Colorado, uranospinite is reported to fluoresce bright yellow-green, with a decrease in brightness due to dehydration.

Meta-uranocircite, $Ba(UO_2)_2(PO_4)_2 \cdot 8H_2O$.

The meta form of uranocircite is apparently the only one found in nature. The mineral is reported from the Wilson mine in Clancy, Montana, and from near the White River area of the Badlands in Pennington County, South Dakota; the fluorescence is reported to be the usual yellow-green.

Heinrichite, $Ba(UO_2)_2(AsO_4)_2 \cdot 10\text{–}12H_2O$.

Heinrichite is reported from near Lakeview, Lake County, California, as a bright fluorescent, presumably yellow-green.

Novacekite, $Mg(UO_2)_2(AsO_4)_2 \cdot 12H_2O$.

Novacekite is found at Laguna, New Mexico, and the Wichita Mountains in Oklahoma. The fluorescence is reported as pale green, pale yellow-green, or yellow. Differences in reported response can be due to substitutions which occur in this mineral, such as phosphorus for arsenic or calcium, copper, or iron for magnesium, or to dehydration which also occurs. Specimens seen by the author fluoresce in the usual yellow-green of most uranium minerals.

Sabugalite, $HAl(UO_2)_4(PO_4)_4 \cdot 16H_2O$.

Reported from the Happy Jack mine, White Canyon, Utah, from Rio Arriba County, New Mexico, and from Cameron, Arizona, sabugalite is usually mixed with other uranium minerals. Fluoresence is reported to be bright yellow.

Metazeunerite, $Cu(UO_2)_2(AsO_4)_2 \cdot 8H_2O$.

Zeunerite occurs naturally in the meta or dehydrated form. Since copper is its primary metallic constituent, this mineral is to be expected under somewhat different circumstances than most of the others of the uranyl bearing group. Thus it is found in rhyolite with torbernite at Majuba Hill in Nevada; with pitchblende at the Wilson mine in Clancy, Montana; in several of the White Canyon and San Rafael mines in Utah; at the Monument No. 2 mine in Arizona; and scattered other places. The fluorescence is reported as weak yellow-green under long wave.

Phosphuranylite, $Ca(UO_2)_3(PO_4)_2(OH)_2 \cdot 6H_2O$.

Phosphuranylite, a peculiar fluorescent mineral, appears on the basis of chemical structure to contain a uranyl group in its composition, but the fluorescence is not the usual shade of green. Rather, the fluorescence is reported as orange-brown, or none. It is found in the Spruce Pine district of North Carolina, and at the Ruggles and Palermo mines in New Hampshire, with meta-autunite. It is also reported from numerous sites in the West which offer other minerals in this list, including mines in the Grants region of New Mexico, as well as from Jim Thorpe, Pennsylvania.

Abernathyite, $K(UO_2)(AsO_4) \cdot 4H_2O$.

Found at the Fuemrol No. 2 mine in Emery County, Utah, abernathyite is reported to fluoresce bright yellow-green.

Uranyl Silicates

Uranophane and Beta-uranophane, $Ca(UO_2)_2Si_2O_7 \cdot 6H_2O$.

The beta form is a dimorph of uranophane, possibly induced by the presence of some lead. The two forms are probably not distinguishable by the collector, just as most of these minerals are difficult to distinguish generally. Uranophane is considered one of the common uranium minerals and is widely distributed; it is reported from most of the locations listed for other fluorescent uranium minerals and from numerous other locations. Uranophane often does not fluoresce, but when it does, the response is reported as weak yellow-green.

Sklodowskite, $Mg(UO_2)_2Si_2O_7 \cdot 6H_2O$.

Reported from the Oyler Tunnel claim and Honeycomb Hills in Utah, and from Jim Thorpe in Pennsylvania, sklodowskite is reported to fluoresce pale yellow, or not at all.

Soddyite, $(UO_2)_5Si_2O_9 \cdot 6H_2O$.

Soddyite is reported from the Jim Thorpe area in Pennsylvania; from the White Oak mine near Nogales, Arizona; from Marysvale, Utah; and from locations in Colorado and Nevada. Its fluorescent response is reported to be the anomalous weak orange-brown found in a few uranium minerals, or none at all.

Boltwoodite, $(H_3O)K(UO_2)(SiO_4) \cdot H_2O$.

Boltwoodite is reported from the San Rafael Swell area, Utah, and in serpentine at Easton, Pennsylvania. The fluorescence of the Easton material is dull green, almost yellow. However, authenticated specimens from overseas fluoresce a dull orange-brown, under long wave ultraviolet only.

Part B: Fluorescent Minerals Worldwide

The fluorescent minerals described in Part A of this chapter, "Fluorescent Minerals of the United States," can also be considered to be a useful guide beyond the borders of the United States. Nature tends to repeat itself, and to a great extent, the fluorescent minerals described earlier can be expected to be found in other mineral localities of the world. Indeed, some fluorescents from outside this country are far better known and far more brilliant than their equivalents here, blue fluorescing fluorites from Durham, England, being an example. Only a very few fluorescent minerals may prove to be so singular, so unique, that they can be found only in the United States.

However, a number of fluorescent minerals are known from outside the United States which may not have been noted within the United States to date. Or, familiar fluorescent minerals may display an unfamiliar type of fluorescence in a specimen found elsewhere. It can be expected that, in the course of time, many of these will be found within the boundaries of this country.

This section describes a number of the fluorescent minerals found elsewhere in the world for those cases in which either the mineral has not been noted as a fluorescent here, or the particular fluorescent color seen is not known in specimens from the United States. It does not need to be said that this list is probably far from complete. The world is vast, the number of reports of fluorescent mineral is limited, and the opportunity to examine specimens from outside this country has been even more limited.

As in Part A of this chapter, the minerals described here are based on firsthand observation, except where the fluorescence is described as "reported." This last will be seen frequently in this section, indicating that the reports of others have been leaned on heavily. In numerous cases, these reports do not give exact localities, and in most cases, it is not clear whether short or long wave ultraviolet fluorescence is described.

All of this represents an opportunity for collectors to refine, correct, and extend the record in the future.

NATIVE ELEMENTS

Carbon variety Diamond. In the Dana system of mineral classification, the elements which are found in nature uncombined with other elements are the first group described. Since those elements which are gases at ordinary temperatures are not considered to be minerals, and since most of the others are chemically reactive and thus rarely seen as minerals in nature, the list of native elements is short. The list of fluorescent native elements is even shorter and consists of one member, carbon in the form of diamond. When a number of diamonds are examined under either short or long wave light, approximately 20 to 30% are found to fluoresce. Colors vary but include blue-white, pink, red, orange, yellow, and green. Diamonds are also known to be phosphorescent and triboluminescent. One of the earliest experiences in the luminescence of minerals was that certain diamonds, held in sunlight for some time and then taken indoors, would glow with enough light to read by. Such diamonds came from India. The source of the cut diamonds whose fluorescence is reported above is unknown, but is probably South Africa.

SULFIDES AND SULFOSALTS

Enargite. Groups of dark gray or black striated crystals identified as enargite have been found in Chantarcillo, Peru, associated with large, cubic, pyrite crystals. Quite contrary to expectation, the mineral fluoresces bright orange under long wave ultraviolet. There is no visible evidence that this is due to an alteration or coating, for example, to sphalerite.

Realgar. Some realgar from Rumania appears to fluoresce yellow, best under long wave ultraviolet. As fluorescence in realgar is quite unexpected, further research is required to confirm that this fluorescence is due to realgar and not to some other material which might be present.

HALIDES

Fluorite. A scarce and exotic fluorite is found as crystals or masses at Mapimi, Mexico. In daylight, the color is the kind of purple typical of much fluorite. Under long wave ultraviolet, the usual blue response is shown. However, under short wave, the fluorite fluoresces an unusual dark red, the color of a red wine. Possibly, samarium is the activator of this fluorescence. An apparently similar red response is reported in certain fluorites from Germany, except that it is reported for both short and long wave ultraviolet.

Marshite. Marshite from Broken Hill, New South Wales, Australia, will sometimes fluoresce a deep dark red under long wave. A similar response is reported for marshite from Chuquicamata, Chile. It appears that marshite (and possibly enargite) are the only copper minerals known to fluoresce.

Prosopite. Prosopite from Greenland sometimes fluoresces blue-white or sky blue under either short or long wave ultraviolet.

CARBONATES

Aragonite. Aragonite crystals from the sulfur mines in western Sicily have long been famous for size, excellence of hexagonal form, and overall display. However, it is as a fluorescent that these aragonites are most spectacular. Under long wave ultraviolet, the fluorescence is a deep rich pink. Under short wave, it is usually white with a greenish tint. On removal of either the long or short wave light, the fluorescence is followed by a strong greenish white phosphorescence.
Carborotite. Carborotite is reported from China as fluorescing white.
Cerussite. Large crystals of cerussite from South West Africa have shown an unusual sky blue fluorescence under long wave ultraviolet.
Gaylussite. Gaylussite from Kenya fluoresces a clear and distinctive blue-white under long wave ultraviolet.
Hydrozincite. Hydrozincite from Mapimi will fluoresce in the usual intense sky blue under short wave ultraviolet, but unlike most other hydrozincite, such specimens may fluoresce orange under long wave ultraviolet.
Mellite. Mellite from Germany and Hungary fluoresces blue-white under short or long wave ultraviolet, with a strong phosphorescence of the same color.
Natrofairchildite. Natrofairchildite from the Soviet Union is reported to fluoresce weak orange.
Otavite. An interesting and rare mineral from South West Africa, otavite can be thought of as a cadmium smithsonite. The color of otavite under either long or short wave ultraviolet is red.
Phosgenite. Large, well-shaped crystals of phosgenite from Sardinia are noted for their form and size. They are equally attractive as a fluorescent. The fluorescent response is yellow to yellow-orange, strongest under long wave.
Strontianite. Strontianite, or possibly strontian aragonite, in clusters of small crystals or radiating masses in gray or gray-green daylight color is found in a quarry in Tuscany, Italy. Under long wave ultraviolet, the fluorescence is an exquisite rich cherry red, quite unusual in this mineral.
Tarnowitzite. Tarnowitzite is a lead bearing aragonite found at Tzumeb, South West Africa. The fluorescence is yellow-white, best under long wave.

BORATES

Ameghinite. Ameghinite from Argentina is reported to fluoresce pale blue.
Fabianite. Fabianite from Germany is reported to fluoresce brownish-yellow,

Johachidolite. Johachidolite is reported from Korea as fluorescing intense blue.
Kurchatovite. From the Soviet Union, kurchatovite is reported to fluoresce bright violet.
Nifontovite. From the Soviet Union, nifontovite is reported to fluoresce violet.
Pentahydroborite. Pentahydroborite from the Soviet Union is reported to fluoresce violet under short wave ultraviolet.
Sassolite. From Italy, sassolite is reported as blue under short wave ultraviolet.
Uralborite. Found in the Soviet Union, uralborite is reported to fluoresce violet under short wave.

SULFATES

Barite. Barite is found with uranocircite in a hematite matrix from Menzen Schwand, Germany. The uranocircite fluoresces green, as expected from a uranyl bearing mineral. The barite fluoresces orange under long wave, an unusual fluorescent color for barite. Barite from Bologna, Italy, possibly the same material whose phosphorescence was first noted by early alchemists, is reported to fluoresce strong orange under both short and long wave ultraviolet.
Lanarkite. From lead deposits in Scotland, the rare mineral lanarkite is reported to fluoresce yellow under long wave ultraviolet.
Orpheite. From Bulgaria, orpheite is reported to fluoresce turquoise blue.
Susannite. A rare mineral from Leadhills, Scotland, susannite is dimorphic with leadhillite. Under either long or short wave, the fluorescence is strong lemon yellow.
Swedenborgite. A rare and interesting mineral, swedenborgite appears to display a very interesting fluorescence. Under short wave, an intense sky blue is seen, while under long wave the response is deep red. Further study of the fluorescence of this mineral is indicated to confirm that these responses are due to swedenborgite and not to an accompanying mineral.

TUNGSTATES

Stolzite. Stolzite, as small crystals from Broken Hill, New South Wales, Australia, appears to fluoresce pale yellow under either short or long wave ultraviolet. Stolzite which fluoresces yellow under short wave has also been reported from a Swedish source. Under the microscope, examination of stolzite from the Property mine at Broken Hill shows that such stolzite crystals are irregularly covered with a yellow-green fluorescent coating which is responsible for the fluorescence usually seen. Under the microscope, the stolzite crystals themselves fluoresce weak orange under long wave ultraviolet and a weaker yellow under short wave.

PHOSPHATES AND ARSENATES

Apatite. A bright glassy blue apatite from the Otter Lake area of Quebec, Canada, fluoresces bright sky blue under short wave ultraviolet.

Clear, colorless apatite from the Morro Velho gold mine in Minas Gerais, Brazil, may fluoresce pink, magenta, or violet–colors difficult to describe but quite beautiful. Sometimes such crystals will show zoning: fluorescing pink, magenta, or violet in some portions and a more conventional orange in other portions. Large, well-formed, clear crystals of apatite from Panasquiera, Portugal, are well known among crystal collectors. These sometimes fluoresce yellow-orange and rarely may also fluoresce violet or magenta under short or long wave ultraviolet. Pink fluorescing apatite has been found with stannite, jeanbandyite, and crandallite in the Contacto vein, Llallagua, Bolivia.

The blue, violet, magenta, and pink fluorescent responses mentioned above are suggestive of rare earth activators. The blue fluorescence in particular is reminiscent of blue responses in fluorite and microcline, and europium may be the activator involved.

Bolivarite. Listed as from Spain and Zaire, bolivarite is reported to fluoresce light green under short wave ultraviolet.

Collinsite. Collinsite from the Flinders Ranges in Australia is reported to fluoresce light green under short wave ultraviolet.

Hedyphane. White coatings of hedyphane from Langban, Sweden, will fluoresce strong orange under long or short wave.

Isokite. Isokite from Zambia is reported to fluoresce blue under long wave.

Jagowerite. Jagowerite from Canada is reported to fluoresce greenish white under long wave ultraviolet.

Mimetite. Mimetite crystals from Tsumeb may fluoresce weak orange under long wave. Mimetite from Broken Hill, New South Wales, Australia, is reported to fluoresce orange under short wave.

Natrophosphate. Natrophosphate is reported from the Soviet Union as fluorescing weak orange.

Pyromorphite. Fine hexagonal crystal groups of pyromorphite from Ems, Germany, are tan in daylight color and fluoresce yellow under long wave ultraviolet. Yellowish fluorescence is reported in green-yellow pyromorphite from Broken Hill, Australia.

Svabite. Specimens of svabite from Ultivis, Sweden, fluoresce intense orange for the most part, but a distinctly pink fluorescense is also evident in some specimens.

SILICATES

Agrellite. Found in a small pegmatite on the Kipawa River in Quebec, agrellite is an exceptionally beautiful fluorescent. The color is a magenta pink, best under long wave ultraviolet. It is often found with dark red fluorescing feldspars or with rare fluorescing vlasovite. Rare earths are present in agrellite and a rare earth is likely to be the activator.

Allophane. Allophane from North Hill, England, fluoresces a rich sky blue under long wave and strong green under short wave ultraviolet.

Beryl. Colorless or white beryl from several sources, including Brazil, Elba, and Norway, fluoresces white or yellow-white under long wave ultraviolet.

Beryllite. This chalky white mineral is found in close association with tugtupite at Ilimaussaq, Greenland. It fluoresces grey-green under short wave only.

Bikitaite. From Bikita, Zimbabwe, Africa, bikitaite – a lithium mineral – fluoresces butter yellow under long and short wave ultraviolet.

Catapleiite. From Sweden, catapleiite fluoresces green under short wave ultraviolet.

Cymrite. From Langban, Sweden, cymrite is reported to fluoresce green or bluish white by different authors; the fluorescent response deserves further investigation.

Danburite. Handsome clusters of white, chisel shaped crystals of danburite, up to 5 inches or more in length, were found at San Luis Potosi, Mexico. Under either short or long wave ultraviolet, these fluoresce bright violet-blue, strongest toward the top. Broken sections of the cluster may fluoresce bright yellow-green, and the combination may be decorated with small bow ties of red fluorescing calcite crystals. The combination provides an exceptionally beautiful fluorescent specimen.

It seems possible that europium is the activator of the violet-blue fluorescence mentioned.

Feldspar. Green, glassy microcline form the Consolidated Zinc mine at Broken Hill, New South Wales, Australia, fluoresces bright white with a slight yellowish tint under short wave ultraviolet. Paracelsian from Wales fluoresces cherry red under short wave ultraviolet.

Gaidonnayite. From Mt. St. Hilaire, Quebec, Canada, the rare mineral, gaidonnayite is reported to fluoresce green.

Garnet. Grossular is found as pure white crystals in the Jeffery quarry, Quebec, Canada. These may fluoresce pink under short wave and orange under long wave ultraviolet. Hydrogrossular from near Mosjoen, Norway, is reported to fluoresce a clear orange under short wave ultraviolet with no response under long wave. A red fluorescing uvarovite has also been mentioned in the literature, in which case chromium can be suspected as the activator. Fluorescing garnet appears to be very uncommon in nature, though fluorescence can be produced in synthetic garnet.

Genthelvite. Genthelvite from Mt. St. Hilaire is reported to fluoresce bright green.

Gyrolite. Gyrolite from Czechoslovakia is reported to fluoresce white.

Hauyne. Hauyne from Italy is reported to fluoresce orange-pink under long wave ultraviolet.

Holtite. Holtite from Australia is reported to fluoresce orange under short wave and yellow under long wave ultraviolet.

Huttonite. Huttonite from New Zealand is reported to fluoresce pink-white.

Leucophanite. This mineral, also called leucophane, is an extremely beautiful fluorescent. Material from Norway or from Mt. St. Hilaire in Quebec may fluoresce brilliant magenta-pink, best under long wave ultraviolet. The fluorescent color is much like that of agrellite, and as difficult to describe. A rare earth is likely to be the activator in this mineral.

Magnesioaxinite. Magnesioaxinite from Australia is reported to fluoresce red under both long and short wave ultraviolet.

Malayaite. An analogue of sphene in which tin acts in place of the titanium in sphene, malayaite from Durham, England, fluoresces bright lemon yellow with a slight green tint under short wave ultraviolet.

Monteregianite. A rare mineral from Mr. St. Hilaire in Quebec, monteregianite is reported to fluoresce green.

Natrolite. The tips of stubby natrolites from Mt. St. Hilaire in Quebec, Canada, may fluoresce green under short wave ultraviolet.

Parakeldyshite. From Langendalen, Norway, parakeldyshite fluoresces weak white under short wave ultraviolet.

Polylithionite. A lithium mica found in Canada, polylithionite may fluoresce tan under short wave.

Quartz. In its noncrystalline form, including hyalite and opal, quartz is a well-known fluorescent mineral, but in the crystalline form, fluorescence under short or long wave ultraviolet in quartz is rare. Thus, the very brilliant short or long wave lemon yellow fluorescence of quartz from Livorno or Pisa, Italy, is quite surprising and startling. Equally unusual is the deep orange-red fluorescence seen in the cores of gray crystals from Otter Lake, Quebec, Canada.

Scapolite. Found as large white blocky crystals at Otter Lake, Quebec, Canada, broken sections of scapolite will fluoresce deep red under short wave and white under long wave ultraviolet.

Sorensenite. From Greenland, the rare mineral sorensenite is reported to fluoresce white or blue-white under long wave ultraviolet.

Spodumene variety Kunzite. From Madagascar and from Laghman, Afghanistan, kunzite may fluoresce pink under long wave and blue-violet under short wave ultraviolet, and may also show an intense red phosphorescence. Kunzite sometimes show dichroism under ordinary light. That is, the color of transparent kunzite may change from pink to colorless when viewed through different directions in the crystal. A similar effect is sometimes evident in the fluorescence of kunzite. The fluorescent color may shift between pink and blue-violet as the crystal is rotated before the ultraviolet light. Further investigation would be required to definitely establish that this is a fluorescent dichroism and not the effect of zoning through the crystal.

Topaz. Topaz will fluoresce rarely. Some topaz from Brazil displays an orange fluorescence under long wave, while some topaz from Norway will fluoresce yellow under long wave.

Tugtupite. A rare mineral from the Julianehaab district, Greenland, tugtupite is one of the most beautiful and sought after of the fluorescent minerals. Chemically, tugtupite has been described as a beryllium sodalite, and like a sodalite (variety hackmanite), it is tenebrescent. Material which is initially white will turn deep pink on exposure to light. Under the long wave light, tugtupite fluoresces a deep rich orange, while under short wave, the fluorescent color is deep cherry red. Occasionally, a blue-white phosphorescence can be seen as in some hackmanite. Tugtupite is sometimes found with fluorescing beryllite.

Vlasovite. Vlasovite, a rare mineral found with agrellite in Quebec, Canada, has a rich yellow-orange fluorescence under short wave which is weaker under long wave.

Vuonnemite. A rare mineral from the Kola Peninsula, U.S.S.R., vuonnemite may fluoresce yellow-white under short wave ultraviolet.

URANIUM MINERALS

Bergenite. Bergenite is reported from Germany as fluorescing orange-brown under both short and long wave ultraviolet.

Calcurmolite. An interesting formulation, being calcium uranyl molybdate, calcurmolite is reported to fluoresce strong green.

Furongite. From China, furongite is reported to fluoresce a strong yellow-green.

Metalodevite. Another interesting formulation, a zinc uranyl arsenate, metalodevite from France is reported to fluoresce yellow-green.

Saleeite. From Australia, saleeite is reported to fluoresce yellow-green.

Strelkinite. From the Soviet Union, strelkinite is reported to fluoresce dirty green.

7
Activators in Fluorescent Minerals

Why does calcite from Franklin, New Jersey, fluoresce red, calcite from elsewhere fluoresce white, green, or blue, while many other calcites do not fluoresce at all? In most cases, the fluorescence is caused by the presence in the mineral of an "activator." An activator is usually thought of as a "foreign" element, sometimes present only in small amounts. When a mineral does not fluoresce, it may be because it does not contain a suitable activator. Of the somewhat more than 90 elements which occur in nature, only a few appear to be important as activators of fluorescence in minerals. These include manganese, uranium, some of the "rare earth" metals, and a few others.

While a great deal is known about activators and their role in producing fluorescence, the activators in a number of fluorescent minerals have not yet been identified. Many foreign elements are found in minerals and it is usually difficult to tell which of these is the activator, particularly since some activators can be effective in exceedingly small amounts. For example, H. W. Leverenz of the RCA Laboratories describes the following experiment in fluorescence.* A few crystals of the mineral sylvite are placed on the edge of a glass plate. A crystal of antimony trichloride is placed on the edge of another plate nearby. Both are illuminated with ultraviolet light. Neither of these materials fluoresces. While still under the ultraviolet light, the two glass plates are moved closer to one another. When they are still one-half inch apart, the sylvite crystals nearest the antimony trichloride begin to fluoresce yellow! Atoms in unbelievably small amounts apparently evaporate from the solid antimony trichloride and reach the sylvite, activating it into fluorescence.

Laboratory experiments of this sort are part of a continuing effort to provide knowledge about the causes of fluorescence and to explore the properties of synthetic fluorescent materials. Much of what is known about the role of activators in fluorescence has been learned in this way. For such study, synthetic or artificial fluorescent materials have an advantage over natural fluorescent minerals in that purity can be carefully controlled and activator elements introduced in known amounts.

This type of work is not carried out purely for the sake of scientific knowledge since artificial fluorescent minerals have very impor-

*Leverenz, H.W. (1968) *An Introduction to the Fluorescence of Solids.* New York: Dover Publishing Co.: page 64.

tant practical applications. When produced for such applications, artificial fluorescent minerals are known as phosphors. The white coating on the inside of a fluorescent light tube is such a phosphor, often synthetic apatite. The screen of the TV picture tube is also a phosphor, and sphalerite- or willemite-type phosphors have been used in color TV. Artificial materials rather than natural fluorescent minerals are used for these practical applications for several reasons. In an artificial phosphor, the activator can be carefully selected, fluorescent color can be carefully controlled, phosphor intensity can be maximized, and uniform phosphors can be produced batch after batch.

HOW ACTIVATORS WORK

Activators can operate in several different ways to produce fluorescence or modify fluorescent response. One group of particularly interesting activators is the "rare earth" elements. These are believed to be responsible for fluorescence in some fluorites, feldspars, and probably also certain calcites. The rare earth elements (and a few related elements) are listed below.

Element	Symbol
scandium	Sc
yttrium	Y
lanthanum	La
cerium	Ce
praseodymium	Pr
neodymium	Nd
promethium	Pm
samarium	Sm
europium	Eu
gadolinium	Gd
terbium	Tb
dysprosium	Dy
holmium	Ho
erbium	Er
thulium	Tm
ytterbium	Yb
lutetium	Lu

These names are not exactly household words, even in the household of a mineral collector! These elements are scattered widely but thinly throughout igneous rocks and in certain minerals in which they may cause fluorescence, but they occur in commercially valuable concentrations in mostly unfamiliar minerals. These minerals include monazite, xenotime, euxenite, gadolinite, and about 160 others, most of which are not well known to the collector. They are rare and seldom available in attractive specimens.

The rare earth elements have some very peculiar chemical properties. While they are distinctive elements, they behave almost identically chemically. This is very unlike the distinctive chemical behavior which usually characterizes different elements. One result of this was that during the nineteenth century, scientists had significant trouble separating one from another and were not at all sure that they were really distinctive. They appeared to be just mysterious variants of one element. Because they are so close chemically, where one of these rare earth elements is found in a fluorescent mineral, others will usually also be present. Further, when several rare earth elements are present, the fluorescence due to one can mask or change the fluorescence due to another. This can make the identification of which rare earth activator is "acting" exceedingly difficult.

In rare earth activators, the fluorescence process is usually completely confined to the atoms of the rare earth. Also, the color of fluorescence differs little among the different minerals in which one of these rare earth elements may be found. Thus, the rare earth element europium produces a blue fluorescence when it is found in several quite distinct minerals, including fluorite. The host mineral is little more than a solvent for the rare earth; that is, it is little more than a carrier and a diluent.

Uranium provides a somewhat similar, self-contained type of activation. The characteristic green uranium fluorescence is widespread in a number of minerals. This really is due not to the uranium atom alone, but to a team composed of a uranium atom with two surrounding oxygen atoms – known as a "uranyl" group. Providing that the mineral encorporates this group in its chemical makeup and crystal structure, the remaining constituents of the mineral have only a small influence on the fluorescence. This is why so many diverse uranium bearing minerals can express a virtually identical "uranium" fluorescence.

In contrast, fluorescence in sphalerite may involve interaction between sphalerite and the activator. In sphalerite, an orange or reddish orange fluorescence is commonly seen. The cause of this fluorescence is not entirely settled, but in one view, it is due to the presence of two activators in trace amounts – copper and indium (or aluminum or gallium). These elements have been found in small quantities in sphalerite and may be the source of its orange fluorescence. Such fluorescence is not produced in pure sphalerite, in copper minerals, or in indium bearing minerals. It is produced only when copper and indium are incorporated in sphalerite in the proper concentration. The fluorescence is thus a result of the interplay between the sphalerite and copper and indium atoms.

Some activators, manganese in particular, are able to generate fluorescent light but are not effective at absorbing ultraviolet. A mineral containing such an activator may not fluoresce. This situation is changed by the presence of a second type of activator, also called a sensitizer. For example, calcite containing only manganese will fluoresce under a beam of electrons but will not fluoresce under short wave ultraviolet. However, if the calcite contains a small amount of lead in addition to manganese, a brilliant red fluorescence results under ultraviolet. In the absence of lead, the ultraviolet presumably passes right through the mineral, or is scattered and lost within the mineral. However, when lead is present, the lead atoms take up the ultraviolet very efficiently and relay the energy to the manganese atoms. The manganese atoms then generate the output of red light. The famous red fluorescence of the calcite from Franklin, New Jersey, owes its brilliance to lead and manganese as dual activators. A number of other well-known fluorescent minerals, including wollastonite and apatite, depend on these dual activators for their fluorescence.

There are other ways in which activators influence fluorescence. Some activator atoms insert themselves between the regular atoms of the mineral and stretch or shrink the distances between those atoms. This sometimes changes the fluorescent capabilities of the mineral. It may allow the mineral to fluoresce more efficiently, or it may change the color of the fluorescence. Often, a number of these roles are fulfilled at the same time by the activator.

Organic matter such as petroleum is occasionally incorporated in minerals and is the source of fluorescence in some of these minerals. The fluorescence of fluorite from Clay Center, Ohio, is sometimes attributed to organic activators.

Water is also thought to play a role as an activator, and it is possible that many blue-white long wave fluorescences are due to a thin film of water wetting the crystal surface, or incorporated water may activate minerals into fluorescence. A review of the fluorescent minerals as described n Chapter 6 will show that a great number of those minerals which fluoresce a pale or weak white contain either water of hydration or the hydroxyl (OH) group in their composition.

Finally, it is important to point out that while activators are thought of as foreign trace elements, in many fluorescent minerals no "foreign" element need be present. As an example, the uranium mineral autunite, $Ca(UO_2)_2(PO_4) \cdot 12H_2O$, fluoresces wonderfully under ultraviolet. The seat of the fluorescence is the uranyl group (UO_2). This uranyl group is not a trace or foreign component in the makeup of the mineral autunite. Rather, it is an intrinsic and necessary component of the mineral. Similarly, no foreign component needs to be present in scheelite to produce the bright and outstanding fluorescence of this mineral. In this case, it appears that tungsten atoms working in coordination with surrounding oxygen atoms are responsible for the fluorescence. Other such cases are known, and the fluorescence of many minerals may ultimately be found to be due to intrinsic elements of the mineral rather than to impurities or trace elements. Such minerals are called self-activated. In a sense, uranium or tungsten could be called the activators in the examples just mentioned since they are the seat or center of the fluorescent response, or with more accuracy, all of the direct participants may be given credit including the oxygen atoms which are essential components of the fluorescent action in these particular minerals. Thus, the activators in these two examples are better said to be "uranyl" (UO_2) and "tungstate" (WO_4) groups.

As can be seen, the question of how activators act is a complicated one, and not all of the questions about the role of activators in fluorescence have been fully answered yet.

ACTIVATORS IN NATURAL MINERALS

Surprisingly little systematic research has been done to determine which activators cause fluorescence in the numerous natural fluorescent minerals. Much of what is known about activators in minerals has been learned through the creation of artificial fluorescent minerals or phosphors. Some of these duplicate natural minerals, and when they fluoresce the same color, it is inferred that the activator in the artificial mineral is at work in the natural mineral also.

Relatively few natural fluorescents have been matched in this way. As a result, the identification of the activator in most minerals of interest to the collector is a guessing game. The starting point would be consideration of the elements which cause fluorescence in artificial materials. These include:

Element	Symbol
rare earths	(R.E.)
titanium	Ti
vanadium	V
chromium	Cr
manganese	Mn
copper	Cu
silver	Ag
tin	Sn
antimony	Sb
tungsten	W
thallium	Tl
lead	Pb
bismuth	Bi
uranium	U

Many of these elements are known to produce fluorescence in minerals. Atoms of manganese, chromium, and various rare earth elements, for example, enter the structure of a number of minerals, occupying the location of some metal which is essential to the formation of the mineral. These intruding atoms are able to replace the native atoms under certain conditions, particularly when the foreign atom is of the same size as the atom replaced. If the electric charge of the two atoms are the same, and if the two link in the same manner to neighboring atoms in the crystal, such replacement is even easier.

Under these circumstances, the introduction of some foreign activator into a host mineral is likely to be a common occurrence. In this way, manganese replaces calcium in a number of minerals, including calcite and apatite, and replaces zinc in willemite and sphalerite, and activates these into fluorescence. Various rare earths are also able to replace calcium. As a consequence, rare earths may be found in low concentrations in calcite, apatite, fluorite, and scheelite. The rare earths may also replace some strontium in strontianite and some lead in pyromophite. Rare earth elements are also found in microcline and other feldspars. In at least some of these cases, the rare earths are known to produce fluorescence and are suspected as the cause in others. Chromium, by contrast, enters only a few minerals as an impurity, replacing atoms of aluminum. Ruby is a particular case of interest, and chromium activates fluorescence in this mineral. Uranium appears in small amounts as an impurity activator in a number of minerals, including calcite and hyalite. It also is an essential component in a large number of other minerals, activating these into fluorescence.

Three activators – manganese, uranium, and the rare earth elements – seem to be responsible for fluorescence in an astonishingly wide range of minerals. The fluorescences which they produce seem particularly brilliant or colorful. If these particular activators were not available to produce their spectacular displays in many minerals, the riches of the fluorescent mineral kingdom would, with a few exceptions, be replaced by a drab gray poverty of fluorescent color.

In Table 7-1, known or suspected activators and the minerals in which they are effective are listed. That comparatively few minerals appear on this list, compared to the much greater number of minerals known to fluoresce, illustrates the present incomplete state of knowledge concerning activators in minerals.

Table 7-1. Mineral Activators.

Activator	Mineral	Fluorescence	Key: (1) certain (2) probable (3) possible
Manganese and lead	anthophyllite	red	(3)
	apatite	orange, yellow	(2)
	axinite	red	(3)
	benstonite	red	(2)
	calcite	red	(1)
	clinohedrite	orange	(3)
	halite	red	(1)
	margarosanite	red	(3)
	pectolite	orange	(3)
	svabite	orange, pink	(2)
	tirodite	red	(3)
	tremolite	orange	(3)
	walstromite	orange, pink	(3)
	wollastonite	orange, yellow	(2)
Manganese (lead as possible co-activator)	esperite	yellow	(3)
	sphalerite	orange	(2)
	willemite	green	(1)
Uranium (uranyl)	(uranium minerals, various)	green	(1)
	adamite	green	(1)
	apophyllite	green	(3)
	aragonite	green	(2)
	calcite	green	(2)
	gyrolite	green	(3)
	hyalite	green	(1)
	monohydrocalcite	green	(3)
Europium	apatite	blue, violet-blue	(3)
	calcite	blue, violet-blue	(2)
	danburite	blue, violet-blue	(3)
	fluorite	blue	(1)
	microcline	blue	(2)

Table 7-1. (continued)

Activator	Mineral	Fluorescence	Key: (1) certain (2) probable (3) possible
Unspecified rare earths	agrellite	pink, magenta	(3)
	apatite	pink, magenta	(3)
	danburite	green	(3)
	leucophanite	pink, magenta	(3)
Yttrium	yttrofluorite	yellow	(2)
Copper (with co-activator in some)	sphalerite	orange, pink	(2)
	sphalerite	blue	(2)
	sphalerite	blue-green, green	(2)
	willemite	yellow-orange	(2)
Silver	sphalerite	blue	(2)
Chromium	corundum	red	(1)
	uvarovite	red	(2)
Mercury	calomel	orange	(2)
Tungsten (tungstate)	scheelite	blue	(1)
Sulfur (polysulfide)	scapolite	yellow	(2)
	sodalite	orange	(2)
Organic	fluorite	yellow-white	(2)

FLUORESCENCE "POISONERS"

Just as some elements cause fluorescence in the role of activators, there are several elements which act to destroy fluorescence in some, perhaps most, minerals. These are known as fluorescence quenchers or poisoners. Three elements are prominent in this role:

Element	Symbol
iron	Fe
cobalt	Co
nickel	Ni

Just as some activators can be effective in amazingly small amounts, these poisoners can be effective in very small amounts. One part in 10,000 of nickel can suppress fluorescence in sphalerite. Since iron occurs in small amounts in many minerals, it is astounding that as many minerals fluoresce as do.

ROLE OF THE HOST MINERAL

Emphasis has been placed thus far on the role of the activator in producing fluorescence, but the host mineral also plays a vital role. As stated earlier, the main or essential elements of the mineral may themselves be the activators, or they may share activation with one or more impurities as in sphalerite. In some cases, the host mineral may act in the place of a sensitizer activator in that it is the primary absorber of ultraviolet. In other cases, they influence the response of the activator in a way which will modify the color of the fluorescent light. However, there is one role which the host mineral plays that is universal to all fluorescent minerals: the host mineral atoms are constantly buffeted by collisions with the activator atoms and, as a result, carry excess energy away from the activator atoms. This is an indispensable activity in fluorescence and will be discussed in the next chapter.

8
Why Minerals Fluoresce

Why is it that some minerals fluoresce and others do not, and that some fluorescent minerals phosphoresce? How does invisible ultraviolet produce this spectacular show of visible light in certain minerals?

It has only been in recent decades that the answers have emerged, that the powerful theoretical tools and experimental methods of modern physics have begun to provide the required explanations. These explanations are derived from quantum theory, a basic and fundamental description of nature found to apply to events involving the atom and its electrons. This theory requires special knowledge which most of us do not have and which is difficult to describe in familiar terms. However, it is still possible to understand the cause of fluorescence in minerals in a general way, in a way sufficient for our needs, providing we are willing to forego the mathematical detail and precision of description which the physicist would require. Such precision is not available in any case for most minerals of interest to fluorescent collectors. Only a small number of minerals – sphalerite, willemite, ruby, halite, scheelite, and a few others – have been studied in any depth, and even in a few of these, some uncertainty still exists with regard to details of the fluorescence process.

AN OVERVIEW

We have all probably read that fluorescence is caused by an electron in an atom being knocked to a higher orbit and that, when it returns, light is given off, or some equivalent explanation, and have been left with an unsatisfied feeling with regard to what is going on. The actual events that take place in fluorescence are significantly more complex and will require an explanation of the interaction of light with the activator and the interaction of the activator with the surrounding material. This interaction involves energy transfers or exchanges between ultraviolet light, activator atoms, and the other atoms of the host mineral. To understand this process, it is necessary to review a few aspects of the physics involved starting with some of the properties of light.

As discussed in Chapter 2, light consists of electromagnetic waves. Light waves carry energy. When light impinges on certain atoms in minerals, the light may be absorbed with a transfer of light energy to these atoms. Atoms may also release light, which will carry energy away from the atom. It is important to understand that the energy transferred by the light during such energy exchanges is related to the wavelength of the light. Short wavelength light carries high energy, while longer wavelengths carry proportionately lower energy. Thus, as shown in Figures 2-2 and 2-3, orange light carries more energy than does red light, simply because orange is of shorter wavelength than red. Similarly, green carries more energy than orange, blue more than green, long wave ultraviolet more than visible blue, and short wave ultraviolet more than long wave ultraviolet.

What is "energy"? Energy is the ability to do work, that is, the ability to move an object against the effect of a force which may oppose the movement. For example, when a weight is lifted, work is done and energy is expended in moving the object in opposition to the force of gravity. When visible or ultraviolet light is absorbed by an atom, the energy of the light is expended in moving electrons against the electrical force which attracts electrons to the nucleus of an atom. The light completely disappears in this process and all of the light energy is taken up by the electron which will now have increased energy. When this happens, the electron can be pictured as transferring from an orbit close to the nucleus to an orbit further away from the nucleus. While this is not always an accurate description of events, it provides a sometimes useful picture.

If the atom is isolated from the company of other atoms, this increase in energy of the electron and of the atom of which it is a part is usually short-lived, and the electron will quickly give up this extra energy and return to its original state, releasing this energy as a pulse of light. For such isolated atoms, the energy given up is equal to the energy initially taken in. What this means is that the light produced when the electron sheds the surplus energy is identical in wavelength to the light absorbed an instant earlier.

Atoms in a rarefied gas are isolated from one another in this way. Mercury atoms in mercury vapor provide an example. If short wave ultraviolet (2537A) is beamed through mercury vapor, the ultraviolet is quickly absorbed as electrons in the mercury atoms shift to higher energy. Equally quickly, they will return to their original energy, and 2537A ultraviolet will be radiated in all directions. The wavelength absorbed and the wavelength radiated are identical, Figure 8-1(a).

This is not quite fluorescence since in fluorescence the light radiated is of longer wavelength than the light absorbed. It is thus necessary to delve further to understand fluorescence.

Up to this point, we have considered the interaction of light with isolated atoms. However, in a mineral, as in other solids, atoms are not isolated from one another. Rather, they are closely packed together. This close proximity produces interference in the energy exchanges just described. One result of this is that, in the atoms of

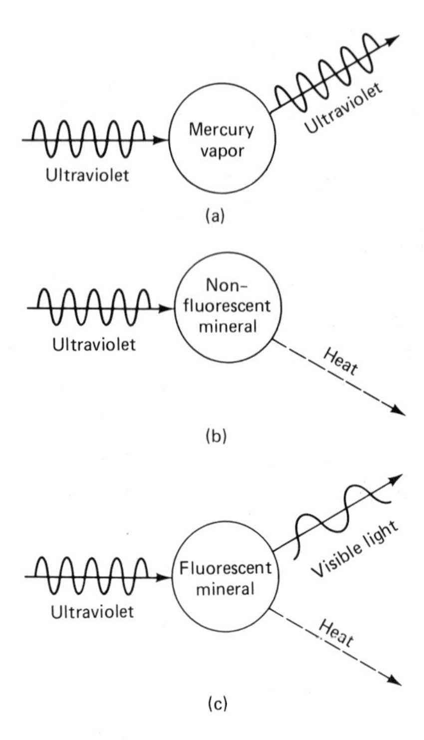

Figure 8-1. Energy taken in when light is absorbed can be disposed of in three different ways. (a) Light absorbed by separate atoms in mercury vapor is reradiated at the same wavelength. (b) Light absorbed in a non-fluorescent material is converted to heat and warms the material. (c) Light absorbed by the fluorescent material is partly converted to heat. The remaining energy is reradiated as light pulses of less energy and therefore of longer wavelength.

most solid materials including minerals, when an electron returns to its original (lower) energy, light is not emitted. Instead, energy is transferred away from the atom by some other means. This is accomplished through collisions with neighboring atoms. The energy originally taken in by absorption of ultraviolet light is now largely transferred to nearby atoms by means of such collisions. These collisions among atoms of a material constitute heat. It is through this means that absorbed light, including ultraviolet, simply warms most materials upon which it falls as shown in Figure 8-1(b).

A fluorescent mineral is an exception, however. In the activator atoms of a fluorescent mineral, some of the absorbed energy, is released internally as heat, much as just described, but the rest is released as a pulse of light somewhat as happens in an isolated atom. Since some of the energy originally received from ultraviolet light has disappeared as heat, leaving only a smaller amount of energy to be released as light, this light will have a longer wavelength than the originally incoming ultraviolet; see Figure 8-1(c). This longer wavelength means that the light produced may well be in the visible portion of the light spectrum. This is the visible light of fluorescence.

STOKES LAW OF FLUORESCENCE

We can see in this behavior an explanation for "Stokes law" of fluorescence. This law states that the light given off in fluorescence is always of longer wavelength than the light taken in to produce the fluorescence. The law is thus a direct consequence of the energy loss within the mineral in the form of heat which always takes place in fluorescence.

It should be stated now that visible light, particularly in the blue and green portions of the spectrum, can also produce fluorescence in some minerals. In accordance with Stokes law, the resulting fluorescent light will be at longer wavelength, typically in the orange or red portion of the spectrum. In fact, this is precisely one of the ways in which fluorescence is produced in ruby. Even red light can produce fluorescence in certain minerals, though in accordance with Stokes law the emitted light, being of longer wavelength than red light, must appear in the infrared portion of the spectrum where it is invisible to human sight. Beryls often fluoresce in the infrared in this way.

To summarize what has been said, fluorescence involves three steps. First, ultraviolet is absorbed in an activator atom and an electron takes on higher energy. Second, the electron begins a spontaneous release of some of this energy as heat, by means of collisions of the activator atom with neighboring atoms. Finally, before all of the energy is released this way, the remaining energy is released in a pulse of light. These are the essential steps in the process that takes place in all fluorescent minerals. With this synopsis in mind, we can now reexamine this process in more detail.

ENERGY BANDS AND FLUORESCENCE

Fluorescence takes place within the atoms of the fluorescing mineral, and so we are concerned with the atom and its outer electrons. The core of any atom, the nucleus, has an intense positive electric charge which binds the negatively charged electrons in various patterns or configurations around the nucleus. These patterns are sometimes called "orbits" but, in fact, are usually far more complicated than the simple circular or elliptical orbits of planetary motion. In any one of these orbits, the electron has a certain total energy. This energy will increase when light is absorbed by the atom of which the electron is a part, and subsequently it may decrease with the release of energy in the form of light or heat. The laws of nature which govern the properties of atoms (and here is where the quantum theory comes into play) dictate that the electron may have only certain specific values, or levels, or orbital energy. When the electron energy changes, it must be from one such energy level to another.

The energy levels can be shown in an energy level diagram. Figure 8-2(a) is a simple illustration for a single isolated atom. In such diagrams, the vertical distance between levels represents the amount of energy difference. A vertical arrow is used to illustrate that the electron has changed energy levels. A long arrow represents a large change of energy, while a shorter arrow shows a smaller change in energy. The arrow pointing upward indicates an increase of energy, which will happen when an atom absorbs light and the electron takes on higher energy. Conversely, an arrow pointing downward indicates that the atom has released energy, either as light or as heat. Such diagrams are frequently used in the discussion of fluorescence processes and will be useful in the discussion to follow.

These levels have the appearance of rungs in a ladder, and to some extent, the analogy is a good one. It is possible to climb a ladder only in fixed steps, and it is necessary to put sufficient energy into the climbing process to reach some higher rung. It is not possible to stand between rungs. Similarly, if light energy is to be absorbed and transferred to the electrons, the light must provide sufficient energy to allow the electrons to reach one of the higher energy levels. As far as most fluorescent minerals are concerned, this usually requires the relatively high energy supplied by ultraviolet light.

THE ROLE OF THE CRYSTAL STRUCTURE

The crystal structure in any mineral involves a regular and repeated pattern or arrangement among the constituent atoms of the mineral. Expanded to billions of atoms in extent, these patterns result in the visible crystal form which is the delight of the crystal collector. Since these repeated patterns of atoms are particularly important in sorting out and classifying the silicate minerals, most books on mineralogy

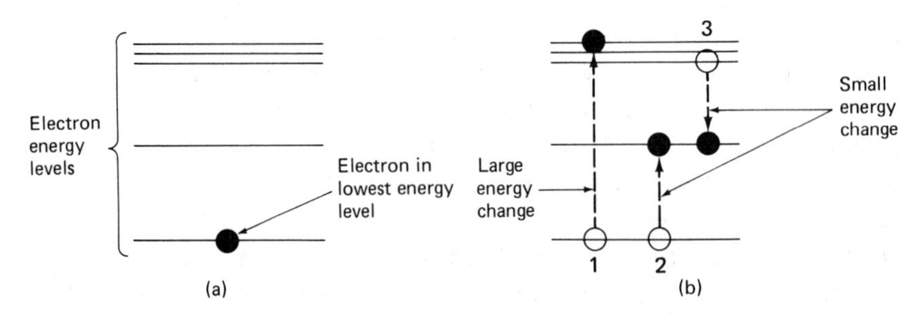

Figure 8-2. In the atom, an electron may change energy in certain allowed steps which can be shown on an energy level diagram. (8-2a). The height of a level above the lowest level is proportional to the energy change. In Fig. 8-2b, 1 shows a large increase of electron energy, 2 shows a small increase, and 3 a small decrease.

will illustrate such atomic patterns, at least for some silicates, and an idea of the complexity and variety of atomic arrangements in the crystal can be gained by consulting such books.

For the present purpose, a highly simplified version of the crystal atomic structure is shown in Figure 8-3, consisting of a single row of atoms. The open circle represents atoms of the host mineral, the filled circle an atom of the activator. These atoms are connected to one another (and to similar other rows of atoms not shown) by electrical forces. These forces generally involve the attraction of the electrons of one atom to the nucleus of the next atom and the counterbalancing repulsion of the electrons of one atom by the other. These forces are elastic, and we can think of the atoms as if they were connected to one another by springs. As a result, the atoms are free to move somewhat and, in fact, are constantly vibrating.

Because of the spring-like coupling between atoms, the movement of one atom is communicated to others, resulting in movement of the others. For example, if we imagine that one of these atoms were to be suddenly pushed to one side, the next nearest atoms would be pushed, and a wave of motion would work its way down the row of atoms. When this happens, energy is transferred from the first atom to the other atoms. By this means, vibrational energy is constantly transferred or exchanged between atoms. These are sometimes called "collisions" but are actually softer pushes through spring-like forces. This vibrational energy, distributed among all the atoms of the material, constitutes the heat content of the material.

While these energy exchanges take place between atoms, the energy gain or loss is transferred to the electrons of the atom, just as energy absorption of light by the atom is transferred to the electrons. This method of energy transfer is not found in the isolated atom, and this introduces significant changes in the energy level diagram of Figure 8-2(a). Due to the close vibrational coupling with other atoms, in place of the single energy levels of that diagram, a group of closely spaced levels is now present representing the small steps or levels of energy exchange due to these vibrations or collisions. Such closely

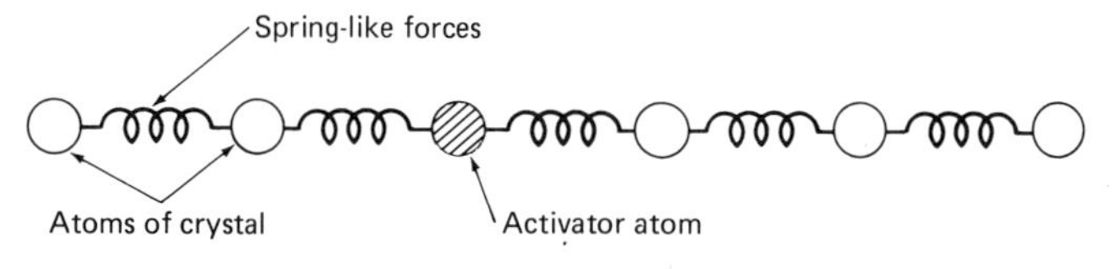

Figure 8-3. The atoms within the crystal structure of a mineral are held in place by a balance of attractive and repulsive electrical forces. In some ways, the atoms acting under these forces behave as if connected by springs, and are able to vibrate and to transfer vibrations down the line of atoms. A simplified model of a row of such atoms is shown above.

spaced levels will be referred to here as a vibration energy band. In a fluorescent mineral, two such bands are usually involved in fluorescence and are shown in Figure 8-4.

If we focus for the moment on one atom within the crystal, the activator atom for example, this atom will be in a greater or lesser condition of vibration depending on whether or not it has recently undergone collision with a neighboring atom. As a result, the activator atom will have greater or lesser energy, as will the electron. This energy will vary from moment to moment. This is shown on the energy diagram of Figure 8-5 by the movement of the electron up and down through the closely spaced energy levels in the band, signifying that the electron is increasing or decreasing in energy. When the atom and the electron with it take up heat energy, the electron moves upward among these energy levels. When the atom gives up heat energy, the electron drops down among these levels.

FLUORESCENCE SUMMARIZED

With this discussion as background, it is possible to describe in further detail the working of fluorescence in minerals. The energy levels of interest for an activator atom in a host mineral are shown in a generalized fashion in Figure 8-5. The electron is usually in the lower vibration energy band, moving up or down within this band as the atom of which it is a part vibrates against neighboring atoms and at various times accepts energy or transfers heat energy to them. However, the energy which may be picked up in this way is rarely ever sufficient to allow the electron to reach the upper vibration band. The energy picked up in such collisions or vibrations is typically 10 to 100 times too small for this to happen. What this means is that vibrational or heat energy is often only one-hundredth the energy of ultraviolet light.

Now, ultraviolet light is directed at the mineral, and some of this reaches the activator atom. The energy carried by ultraviolet is sufficient to transfer an electron to the upper vibration band, as shown in Figure 8-6(a). The ultraviolet is absorbed and its energy completely

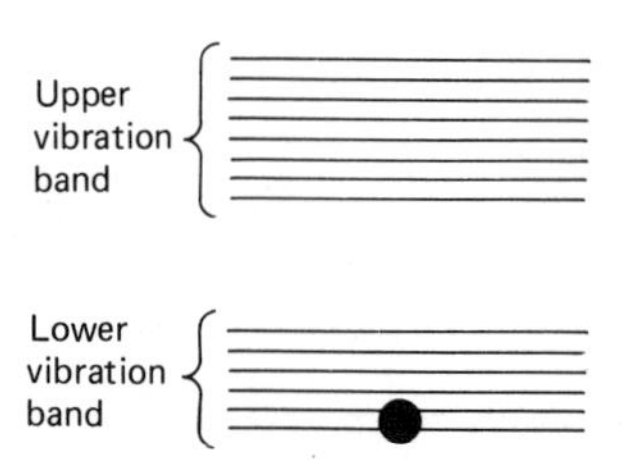

Figure 8-4. In a solid material, the energy levels are grouped into bands of closely spaced levels. In fluorescence, two such bands separated by a gap are usually involved.

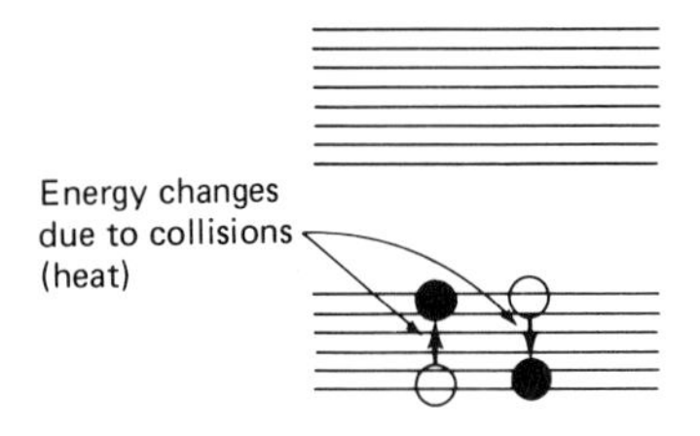

Figure 8-5. Collisions of vibrational exchanges with nearby atoms (heat) change the energy of the electron in the atom by small amounts, represented as movement of the electron among the close levels of the lower band.

converted to the use of the electron which is now in a highly energetic state. Once in this state, the change in the orbit or configuration of this electron around its nucleus changes the equilibrium of the atom in the electric field of its neighbors. It is as if a push had been suddenly delivered to the activator atom in Figure 8-3. As a result, the atom is thrown into increased vibration. This vibration is soon conveyed to the other atoms and energy is transferred to them. This energy is given up by the electron in the upper band of the activator atom. It gives up this energy in small steps, dropping down through the closely packed energy levels of the upper band until it reaches the bottom level of this band, as shown in Figure 8-6(b). As it does this, the highly energetic electron has given up much of its new-found energy as heat to neighboring atoms.

Now, with the electron at the lower edge of the upper band, a new event takes place. The electron gives up all, or almost all, of the remaining surplus energy picked up from the ultraviolet radiation and, in so doing, drops into the lower energy band; this is shown in Figure 8-6(c). The electron gives up this energy in one burst, in the form of light. This light radiates away from the activator atom and leaves the

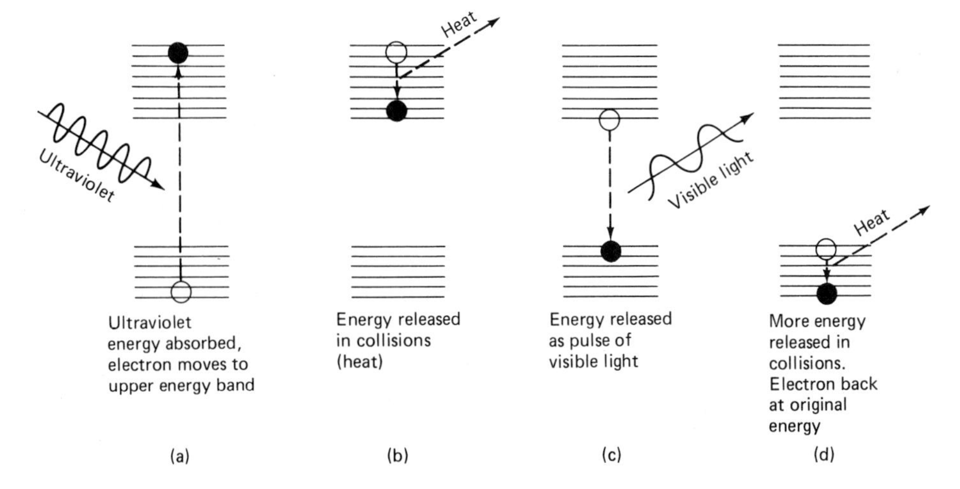

Figure 8-6. Fluorescence typically involves a cycle of four steps. Actual cases will differ, mineral to mineral, but will usually involve some variation of the processes described above.

mineral. This is the visible light of fluorescence. The light produced in this process is of lower energy than that taken in earlier from the absorbed ultraviolet, since some of that energy was given up to other atoms as heat. Being of lower energy, the emitted light is of longer wavelength, sufficiently low that it is in the visible portion of the light spectrum.

The electron has now shed all, or almost all, of the extra energy absorbed from the ultraviolet earlier. It is thus now back in the lower energy band where it started. Here, it may give up further energy in the form of additional heat as it drops within the lower band to a stable energy level, as shown in Figure 8-6(d).

The electron, back in the lower band, is now ready to repeat the cycle when another pulse of ultraviolet reaches it. Meanwhile, billions of other activator atoms are going through the same process, thus producing the intensity and volume of light which can be seen by the eye. Meanwhile also, the mineral is slowly warming, since a certain portion of the incoming ultraviolet is converted to heat, as described earlier.

This is the general principle by which fluorescence is produced in minerals, but in each fluorescent mineral, the principle is worked out in some different or unique way. Just as one mineral is different from another in appearance, chemistry, and physical properties, the fluorescent process will also differ in detail from one fluorescent mineral to another. In many, perhaps most cases, the exact details are not known. Insufficient research has been applied, or even when extensive study has been carried out, the facts developed are subject to different interpretations. Thus, decades of research into the fluorescence of willemite and sphalerite have not yet fully settled the details of the fluorescence in these minerals. However, by using what is known, some of the various and wonderful workings of fluorescence in some specific minerals can be sketched out.

EXAMPLES

Fluorescence in Ruby

Some collectors are fortunate enough to have specimens of the lightly colored pink corundum from the Franklin marble in the vicinity of Franklin, New Jersey, or deeper red corundum, true ruby, from Norway or India. Such specimens fluoresce deep velvet red under long wave ultraviolet. The cause of the red fluorescence of ruby is better understood than fluorescence in most minerals, partly because its processes are comparatively simple and partly because such knowledge was essential in the development of the first laser.

The fluorescence of ruby is due to atoms of chromium which take the place of some of the aluminum atoms in corundum, typically from a few percent to a fraction of one percent in concentration. The electric field produced by surrounding oxygen atoms produces in these chromium atoms a series of electron energy levels which play an important role in fluorescence. Some of these levels are narrow, and some are broad bands spanning a range of energy values. A simplified version of the electron energy level diagram of the chromium atom in corundum is shown in Figure 8-7.

Three energy levels are shown in this figure. The lowest energy level is that in which the electron is usually found in the absence of incoming light. This level is narrow, essentially a single level. The upper energy band is a series of closely spaced levels resulting from interaction of the chromium atom with the adjoining oxygen atoms of the corundum crystal. In the middle, there is an energy level which is broadened to a lesser degree.

When long wave ultraviolet impinges on the atom of chromium, sufficient energy is imparted to the electron to enable it to assume the higher energy level of the top of the upper band. This is shown as (1) in Figure 8-7. In short order, the change of equilibrium of the chromium atom induces vibrations which discharge energy to nearby atoms. The electron loses energy (2), finally reaching the bottom of the middle band. Then, in one massive release of energy as a pulse of red light, the electron returns to the lower level (3). This is the red light of ruby fluorescence.

The wavelength of the fluorescent light is approximately 6950A, close to the extreme red end of the visible light spectrum. Also, the final two energy levels involved in step (3) are very narrow, and as a result, the red light produced occupies a narrow range of energies and

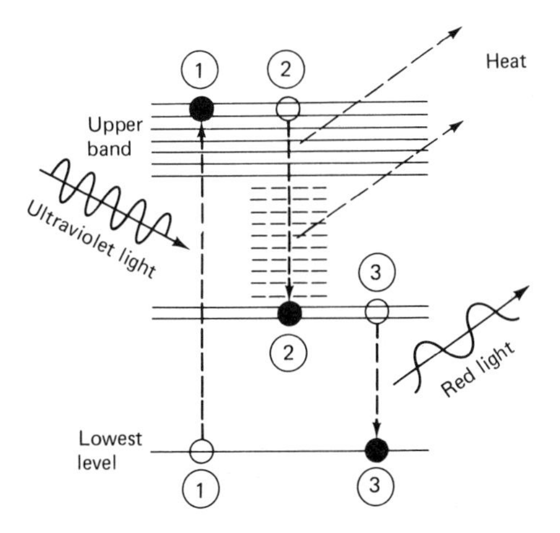

Figure 8-7 A simplified energy diagram of ruby. Visible yellow, green, blue, and long-wave ultraviolet can move the electron to an upper band. For simplicity, only the one upper band is shown, the one to which an electron can be elevated by long-wave ultraviolet. Once ultraviolet is absorbed and the electron is in the upper band (1) vibrations will reduce electron energy (2) until the electron reaches the middle level. From the middle level, the electron sheds remaining energy by emission of a pulse of red light (3) and returns to its original energy.

so is very pure (monochromatic). For both of these reasons, the fluorescence of ruby appears as a very deep red.

If we return our attention to step (1) of Figure 8-7, visible blue or violet light will not have sufficient energy to shift the electron to the top of the upper band but can provide sufficient energy to shift the electron to the bottom edge of this same upper band. From there, the electron will shed energy as described earlier. As a result, visible blue or violet light can also produce red fluorescence in ruby. Further, in ruby an additional band exists in the region between the upper and middle bands of the figure. For simplicity, that band has not been shown. The omitted band is one into which the electron can be moved by visible yellow and green light. Once the electron is in that band, it will follow a course similar to steps (2) and (3); thus, yellow and green light can also cause red fluorescence in ruby. These wavelengths of light – ultraviolet, violet, blue, green, and yellow – can thus produce the red fluorescence of ruby. This fluorescence is one of the reasons ruby is red.

However, there is a second contribution to the red color of ruby. It should be noted that if white light is directed at ruby, the visible violet, blue, green, and yellow components are absorbed to produce the electron energy changes just described. The subtraction of these wavelengths from white light passing through the ruby leaves red. Reflected from some inner surface, the red light is returned to the eye. Thus, both fluorescence and absorption together contribute to the red color of this gemstone.

The fluorescence process in ruby – the absorption of light and the generation of red fluorescent light – takes place in the chromium atom. Similar self-contained fluorescence often takes place in rare earth activators. The blue europium fluorescence in fluorite is probably this self-contained kind.

The Ruby Laser

There is an intimate relationship between the operation of a laser and the ability of the material used to fluoresce. In fact, the first laser utilized synthetic ruby as the laser material.

A laser is a device which produces light in the manner of a fluorescent material, by means of electrons shifting between two energy levels. Unlike an ordinary fluorescent, however, the laser produces a narrow light beam of highly concentrated energy, produces this light in a very narrow spectrum of high purity of color, and maintains a very smooth wave front of light. The result of this high degree of control in the laser has allowed it to be used in innumerable applications including optical range finders, light scanners for high speed printers, surgical cutting devices, and others.

How does a laser work and how is this high degree of control in the laser accomplished? In a fluorescent material, each activator atom discharges a pulse of light at some time of its own choosing, not related to the time of release in some neighboring activator atom. By contrast, in a laser all light producing activator atoms are made to act in synchronism. To understand how this is accomplished, one additional fact must be developed which was not required in the earlier description of fluorescence.

Figure 8-8(a) shows two energy levels of some idealized, isolated atom. We will assume that the energy difference corresponds to the energy supplied by red light. In accordance with explanations given earlier, if red light is absorbed by this atom, the electron will shift from the lower energy level to the higher level. In its own good time, the electron will return to the lower energy level, producing a pulse of red light in the process.*

Now, an interesting phenomenon may occur. If the electron is in the upper level and red light continues to be directed at the atom, as shown in Figure 8-8(b), the electron will be forced back to the lower energy level. This will result in a pulse of red light, but the light wave produced will be in step, wave for wave, with the incoming forcing light. Both waves will proceed onward in step to the next activator atom. Here, if there is an electron in the upper level, another light wave pulse joins the first two (Figure 8-9). The waves will sweep onward creating an avalanche of in-step light as additional atoms of activator throughout the material are triggered. This is the essential activity in the laser.

This effect is supported further if the laser is constructed with flat, parallel ends. These ends are silvered to make two inward facing mirrors. When the light wave reaches either end, it is reflected back into the ruby, sweeping up yet more atoms in its light triggering actions. The reflection back and forth, working in conjunction with the amplification of light intensity produced at each sweep, produces the

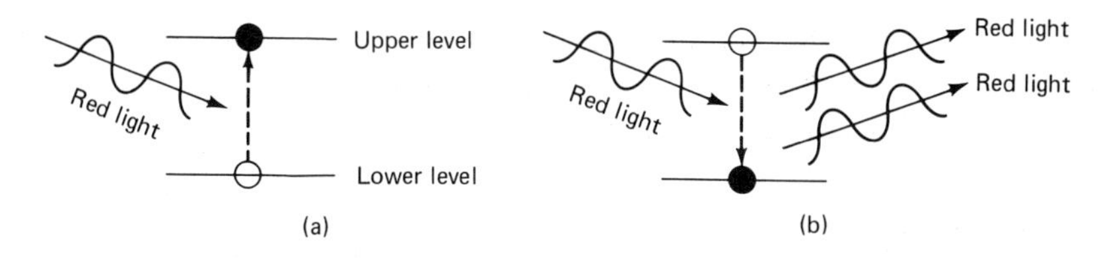

Figure 8-8. Red light is absorbed. The electron moves to a higher energy level. Further red light forces the electron back to the lower level. This light is not absorbed but passes on, joined by the new light pulse.

---

*In the earlier discussion of mercury vapor, short wave (2537A) was absorbed and then reradiated. The same behavior is described here for two energy levels in chromium for which the energy separation between levels corresponds to red light, not ultraviolet.

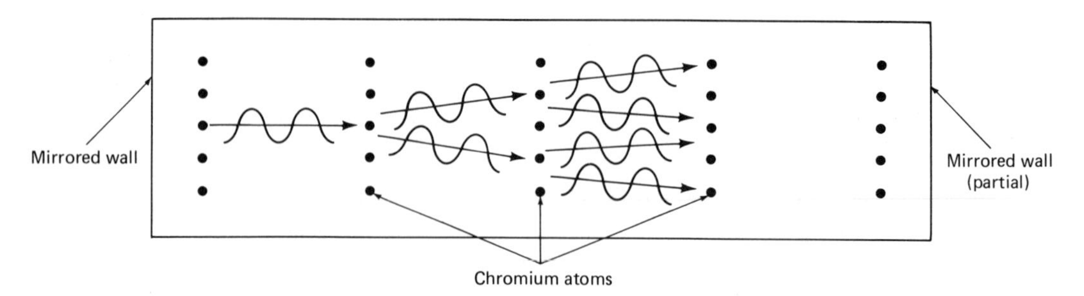

Figure 8-9. A cylinder of ruby laser material. A first pulse of red light will trigger the production of other pulses in synchronism. Light intensity grows rapidly as the light front sweeps forward.

intense beam and narrow spectrum. To allow the light generated in this way to exit the laser for external use, one mirror is only partially silvered so that it is also partially transparent.

Now to consider the role of ruby as the first laser material used: if electrons of the activator atoms are initially in the lower level, they must first be brought to the higher level. How is this to be done? If red light is directed at a mass of atoms with only two energy levels as shown in Figure 8-8(a), electrons will move to the higher level, but soon, the same red light will begin to force these electrons back to the lower level. The second step produces light, but the first step absorbs light. The net result is for one effect to cancel the other. What must be done is to provide some other means to move the electrons to the second energy level, and here is where the three levels of ruby come into play. To return to the energy level diagram of ruby in Figure 8-7: intense violet or ultraviolet light is directed at the ruby. This moves most electrons from the lower or first level to the third level. From here, they return to the second level as described in the discussion of fluorescence in ruby, each atom at its own time. In ruby, the speed with which electrons depart on their own from level three to level two is fast compared to the speed of descent from level two to level one. This allows electrons to accumulate in level two. Now, with a high concentration of electrons in level two, some stray red light, produced perhaps by some fluorescing chromium atom, will trigger the avalanche of light discharges in the other chromium atoms and a laser flash will be produced.

Fluorescence in Scheelite

In ruby, fluorescence is confined to the atoms of chromium. Ultraviolet is absorbed and red light is produced within the chromium atoms. In scheelite by contrast, pairs of atoms are involved. This pair is made up of a tungsten atom and a neighboring oxygen atom. In the crystal structure of scheelite, each tungsten atom has given up electrons to surrounding oxygen atoms. The electron spends most of its time in the vicinity of the oxygen atom, which indicates that a lower energy level is available in the vicinity of the oxygen atom. This level is equivalent to the lower energy band of Figure 8-6. Now, if sufficient energy is imparted to this electron from outside, it can be temporarily driven back to the tungsten atom. The tungsten atom can be viewed as having a higher energy band equivalent to the upper band in Figure 8-6.

When short wave ultraviolet is directed at scheelite, it may be absorbed at the oxygen-tungsten pair which can be considered to be the activator. The electron is driven back to the tungsten atom in an energy change much as in Figure 8-6(a). Thereafter, energy is lost through vibrations. The electron produces a pulse of blue light, much as in Figure 8-6(c), as it returns to the oxygen atom. This process, in which the electron is driven from one atom to a nearby one, is called charge transfer.

Charge transfer need not be limited to atoms of dissimilar elements or to atoms which make up the intrinsic structure of the mineral as is the case with oxygen and tungsten in scheelite. To provide an example, a digression to another interesting area is in order – the daylight color of minerals. There are several different causes for the many varied colors of minerals. In the case of ruby, both light absorption and fluorescence are involved. In some other cases, charge transfer is involved. When the movement of the electron from one atom to another is forced by visible light rather than ultraviolet, those visible wavelengths which produce the electron transfer or charge shift are absorbed in the process. Subtracted from incoming white light, this absorption leaves the remaining light colored. Thus, the mineral appears colored. In the case of the mineral vivianite which is occasionally blue, this color is caused by charge transfer between iron atoms in pairs. One of the pair has given up two electrons to surrounding atoms; the other of the pair has given up three. Visible light can then cause charge transfer between the two atoms of such a pair, from the iron atom with more to the iron atom with fewer electrons. The essential point is that atoms of the same element are involved. It is likely that such charge transfer involving identical atoms can take place due to ultraviolet and that, as a consequence, fluorescence can result. Thus, when it is suggested that paired mercury atoms in calomel or paired sulfur atoms in scapolite are the cause of fluorescence, it is quite possible that it is a charge transfer type of fluorescence.

Fluorescence in Calcite, Apatite, Halite, Wollastonite

Manganese is found as an activator in numerous minerals of widespread occurrence, causing the red fluorescence in calcite, for example. However, manganese is a poor absorber of ultraviolet. In most cases in which manganese produces mineral fluorescence, it is aided by some coactivator. Usually, this coactivator is lead.

The lead atoms, like the manganese atoms, are usually present in the host mineral in small amounts. The lead atoms are very effective in capturing ultraviolet, particularly short wave, and if it were not for the presence of nearby manganese, the ultraviolet taken in would soon be given off as pulses of ultraviolet light by the lead atoms. When manganese is present, the manganese atoms are capable of "seizing" the surplus energy from the lead atom before the lead atom has a chance to release this energy as ultraviolet light. This can happen even though the lead atom is separated from the nearest manganese atom by a few dozen atomic spacings. The process involved is called "resonant transfer." It can take place when there is a close match between the energy to be released from the lead atom and the energy needed by the electron in the manganese atom in order to reach the upper band. This energy transfer process is diagrammed in Figure 8-10. First, the electron in lead moves to the upper or high energy band by absorbing ultraviolet. Now, as shown in Figure 8-10(a), the electron in the lead atom drops back to the lower energy level in the lead atom. As it does so, the released energy is seized by the electron in the manganese atom which moves to the upper band, as shown in Figure 8-10(b). As a consequence, the manganese atom is now ready to emit visible fluorescent light. In this resonant transfer process, energy is carried from the lead to the manganese atom by means of an electric field pulse transferred between these atoms.

Resonant energy transfer involving lead and manganese is important in the red, orange, or yellow fluorescence of calcite, wollastonite, apatite, halite, and probably numerous other mineral fluorescences. Resonant transfer involving other activators is probably involved in yet other mineral fluorescences.

Fluorescence in Willemite

Manganese is the activator of green fluorescence in willemite, just as it is the activator of red fluorescence in calcite and other minerals.

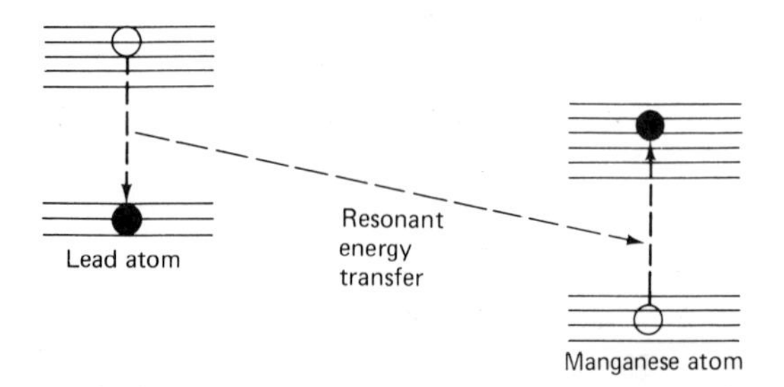

Figure 8-10. As the electron in lead is releasing energy, an electric field pulse is generated which reaches the manganese atom. This atom is tuned to this amount of energy and absorbed it. An electron in the manganese atom is shifted to a higher energy level as a result.

The difference in fluorescent color is due to differences between the two minerals in the deployment of oxygen atoms surrounding the manganese atom and to resulting differences in the electric field imposed on the manganese atom. Such differences are capable of distorting energy levels in manganese, and thus of changing the energy and wavelength of the fluorescent light produced. This type of effect on the energy levels of an atom due to the electric field created by surrounding atoms is called a crystal field effect. The creation of new, separate energy levels in atoms of chromium by the electric field of surrounding atoms in ruby, mentioned earlier, is another example.

In willemite, manganese does not require a coactivator such as lead in order to capture short wave ultraviolet. Rather, the willemite operates in collaboration with manganese to capture ultraviolet energy and to produce fluorescence. In order to see how this is done according to one view, it is first necessary to consider energy levels in willemite.

If light of sufficiently high energy is directed at willemite, electrons may be torn completely free of their parent atom. Once free of the parent atom, the electron may wander freely along selected paths through the crystal until it is recaptured by its original parent or by another atom similarly deprived of an electron. The energy level attained by the electron when it is able to wander freely is called the conduction band since the presence of freed electrons renders the crystal electrically conductive (and since a band of close energy levels is involved). To represent the fact that the electron, once in this band, can drift from the vicinity of one atom to another, the band is shown in Figure 8-11 spanning many atoms of the host ineral. When manganese is present in small amounts in the willemite crystal, its energy levels are imposed locally over those of the host crystal, as shown.

What happens when ultraviolet light is directed at willemite? If short wave (2537A) ultraviolet is used, electrons in the lower energy band

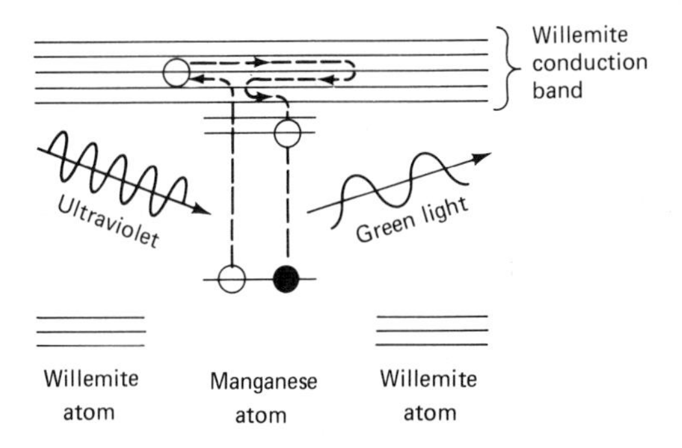

Figure 8-11. Electrons freed of their parent atom can wander through the crystal with the enhanced energy represented by the conduction band. Short-wave ultraviolet will propel a manganese electron into this band. The electron will be recaptured by the manganese atom, give up energy as heat, and produce a pulse of green light.

of willemite will not be provided with sufficient energy to reach the willemite conduction band but will remain in the close vicinity of their parent atoms. In willemite, it would require an ultraviolet wave-length shorter than 2000A to provide sufficient energy to free electrons into the conduction band. However, in the case of manganese atoms in place of some zinc atoms in willemite, the electron is less strongly bound to the manganese atom and it requires less energy to break this electron free. This is indicated in Figure 8-11 by the fact that the lowest manganese energy level is closer to the conduction band than are the lower levels in willemite. Short wave (2537A) ultraviolet will provide sufficient energy to do this and will raise the manganese electron into the willemite conduction band. Thus, while manganese alone will not absorb short wave ultraviolet, manganese embedded in willemite will. It accomplishes this by collaboration between the lowest energy level in manganese and the willemite conduction band.

This first step is equivalent to step (a) of Figure 8-6. Once in the higher band (the conduction band), the freed electron may wander around through the willemite crystal, but in a short time, it will be recaptured by the manganese atom or by some other manganese atom similarly stripped of an electron. In the manganese atom, the electron will first discharge some energy as heat as it drops through the upper energy levels of manganese. Then it will drop to the lower level and emit a pulse of green light, thus carrying out steps similar to (b) through (d) of Figure 8-6. The fact that the electron is, for a time, free of the parent atom and able to drift through the crystal plays an important role in willemite phosphorescence, which will now be discussed.

PHOSPHORESCENCE

One of the intriguing aspects of the behavior of mineral fluorescence is the afterglow, the continued radiation of fluorescent light from a mineral even after the ultraviolet light is removed. Probably all fluorescent minerals have such an afterglow, or phosphorescence, though in most minerals the duration may be too short to be directly and easily seen. In many cases, however, phosphorescence is seen to last many seconds or minutes. In fluorite from Trumbull, Connecticut, eyes adapted to the dark can easily detect continued phosphorescence for many hours after the ultraviolet is removed. On the other hand, some phosphorescences are seen only as a momentary burst of light, or flash, in the instant that the lamp is snatched away from the specimen.

In 1950, the Millsons, father and son, reported the results of a remarkable series of experiments in phosphorescence.* They found that the dark-adapted eye could still detect phosphorescence in Terlingua calcite over 4000 hours after exposure to ultraviolet, in fluorite from Gillete quarry almost 5000 hours after exposure, in certain Canadian scapolite (wernerite) 4,500 hours after exposure, and in the Trumbull fluorite mentioned above an amazing 36,000 hours after exposure. Then, using the sensitive light detecting properties of a photographic plate, they were able to detect phosphorescence in Terlingua calcite some six years after initial exposure.

It is, of course, as interesting to understand the cause of phosphorescence as it is to understand the cause of basic fluorescence discussed earlier. Unfortunately, less is known about the specific mechanisms at work which cause phosphorescence than is known about fluorescence. As a result, only a broad outline of a few possible mechanisns is presented here.

In Figure 8-6(c), the electron shown gives up surplus energy in the form of a pulse of visible light. This ordinarily happens a very short time after the initial ultraviolet excitation. For phosphorescence to occur, something in the fluorescence process must impede or delay the electron in its release of light energy. When this happens, various activator atoms scattered throughout the mineral produce light at different times, some earlier and some later. This spreads the total output of light over some time after the original ultraviolet is removed. These delays may stretch out over years in the case of some minerals.

What is required is an explanation of how this can occur; that is, how the electron can be inhibited for some time from giving up its surplus energy in the form of visible light. Several different causes are believed to be at play in one or another mineral.

However, it is likely that a distant "trap" plays a role in each case. A trap is a location in the crystal where some form of imperfection will seize an electron which has been wandering through the crystal. Crystal defects, missing atoms, or substitution of some foreign atoms in the crystal can trap electrons in this way. Once in the trap, electrons must await chance events in order to be freed. These events are provided by the vibrations of the atoms in the vicinity of the trapping site. When these vibrations are sufficiently strong, the electron may be freed from the trap. Such atomic vibrations constitute heat within the mineral. Thus, increased temperature will more rapidly free such electrons. Once free, the electron may return to the site of fluorescence and proceed to emit light.

---

*Millson, H.E. and Millson, E.M., Jr. 1950. Observations on exceptional duration of mineral phosphorescence. *J. Opt. Soc. Am.* **40** (7):430 - 435.

Since each trap is released by chance events at some different time, the return of various electrons to some site of fluorescence is spread out over time, and thus the production of light is stretched out. This process can be seen at work in the case of willemite.

Phosphorescence in Willemite

Green fluorescing willemites generally show a sustained phosphorescence, or continuous glow, after the ultraviolet light is removed. The duration and the intensity of this glow will vary from specimen to specimen. It seems to be strongest and most persistent in white willemite from Franklin, New Jersey. This may be a secondary or redeposited willemite and may be particularly free of iron, a notorious killer of phosphorescence. The phosphorescence of this willemite may be due to the presence of arsenic, known to produce a phosphorescence of long duration in synthetic willemite. The exact way in which arsenic does this is uncertain, but in some manner, it produces traps scattered throughout the willemite. Arsenic in small amounts is certainly not lacking at Franklin and the nearby Sterling mine at Ogdensburg, and only a small amount is needed.

Figure 8-12 is similar to Figure 8-11 used in the discussion of fluorescence in willemite. The diagram has been extended to the right to permit the presence of a trap distant from the location of the manganese atom to be illustrated. Such traps have energy levels of their own, illustrated very schematically. An electron, freed from a manganese atom by ultraviolet, may wander through the crystal, drift into the vicinity of the trap, and fall into the trap. It will give up energy as heat in doing so. Now to free itself from the trap, sufficient energy must be resupplied to raise the electron back up the steps of the trap. Heat energy, expressed as atomic vibrations, usually accomplishes this. In Figure 8-12, the electron has been raised to the highest energy level of the trap by heat energy relayed by vibrations.

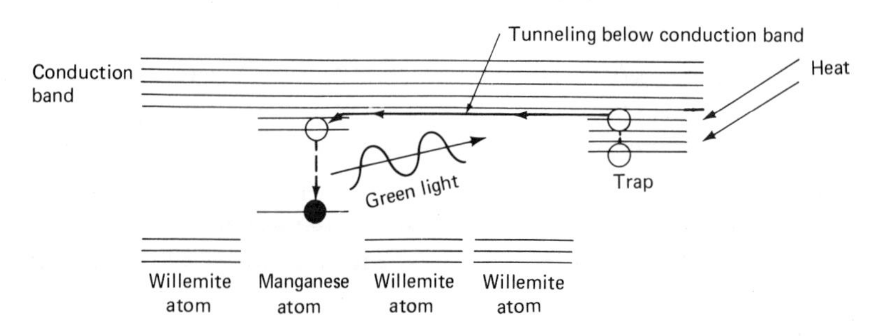

Figure 8-12. Phosphorescence in willemite. The electron, freed from manganese and drifting through the crystal, may encounter a trap. It will sink to low energy levels of the trap. Heat will raise the electron to the top of the trap. From here, the electron will return to the manganese atom by tunneling. Long delays of electrons in the traps produce phosphorescence.

Once in this level, if further heat energy is supplied, the electron may spring free, enter the conduction band, and from here, wander back to the manganese atom. In some phosphorescent minerals, that is precisely the escape route. However, in willemite, at ordinary temperatures, sufficient heat energy to escape out of the top level of the trap is generally not available. What does happen is a process of escape which can occur only on the unbelievably small scale of atoms and electrons. The electron near the top trapping level "leaks" back to the manganese atom directly from the trap. This process is called tunneling. It can be accomplished if the trap and the manganese atom are not too far apart, perhaps a half dozen or so atomic spacings in the crystal. Once back in the upper energy levels of manganese, the electron produces fluorescence as discussed earlier.

THERMOLUMINESCENCE

If a small chip of chlorophane fluorite is placed on the heating element of an electric stove and heated in the dark, a brilliant turquoise green light will begin to glow forth from the mineral as the temperature rises. This may continue for a number of minutes and then cease. Other fluorescent minerals will do much the same, each producing some characteristic color, beginning at some temperature peculiar to each particular mineral. Even some minerals which are not fluorescent under ultraviolet will produce a burst of light if heated sufficiently. This process, known as thermoluminescence, is very closely related to phosphorescence.

Returning to Figure 8-12, we can ask what would happen if the lowest energy levels of the trap were very deep down below the conduction band. Under such circumstances, once an electron becomes trapped, it may stay trapped forever since available heat energy is simply inadequate to raise the electron over this large energy range. However, if the specimen is heated increasingly intense vibrations will run up and down the lattice of atoms, and such energy will find its way to the trap. Then, if a sufficient temperature is reached, traps throughout the mineral will discharge over a very short period of time. The electrons, now free, will rapidly encounter a fluorescence activator site and produce the visible light as in fluorescence. This temperature stimulated production of light is called thermoluminescence. In this sense, phosphorescence is a thermoluminescence in which, for that mineral, ordinary temperatures are high enough to discharge traps.

NONFLUORESCENT MINERALS

At this point, we may ask, "Why don't all minerals fluoresce?" There are a great number of reasons for this, and as a result, it is not so surprising that fluorescence is fairly uncommon among minerals. A few of these reasons are as follows.

First, ultraviolet does not penetrate into some minerals. This is true of the native metals, and is also probably true of those oxides and sulfides which have a metallic appearance in visible light.

A second group of minerals does not fluoresce because the excited electron drops from the upper band to the lower band without radiating a pulse of visible light in the process. Rather, conditions in the mineral are such that the energy gap between the upper and lower bands is bridged with vibration energy levels. This means that the excited electron gives up all of its surplus energy as heat, that is, in collisions with neighboring atoms. This is simply the process of ordinary light absorption in minerals, as in other substances, in which all of the absorbed light is converted to heat. This is the reason that so few dark or highly colored minerals fluoresce. In some minerals, fluorescence "poisoners" are present and are sometimes effective in very small concentrations. These poisoner atoms are capable of robbing an activator atom of its surplus energy obtained from incoming ultraviolet light. The poisoner atom then converts this energy completely to heat, much as described above.

A third group of minerals may not fluoresce simply because the energy separation between the lower band and the upper band of the atoms of the mineral, or of any potential activators present, is so great that even ultraviolet does not provide sufficient energy to raise the electron to the upper band. Such minerals thus do not absorb ultraviolet. They are transparent to ultraviolet light. This is probably the reason that most minerals, particularly those of a transparent or white appearance, fail to fluoresce.

Finally, some minerals may fluoresce, but the light which is produced may be in the ultraviolet or the infrared portion of the light spectrum, and thus invisible. Some demantoid garnet, greenockite, and topaz, among others, fluoresce in the infrared, sometimes with visible light rather than ultraviolet as the input stimulus to fluorescence. It is likely that some minerals fluoresce in the invisible ultraviolet also.

9
Seeing in Color

The most brilliant colors to be seen are the spectral colors, that is, the colors of the rainbow or those seen when white light is viewed through a prism or through a spectroscope. Anyone who has not done so should seek out some opportunity to view sunlight through a spectroscope to appreciate the brilliance and variety of these colors.

The phrase "all the colors of the rainbow" is often used to express the idea of the fullest range of colors. The truth is, however, that while the spectral colors are the most brilliant, they are far from the total number which can be perceived and discriminated by the human eye.

We need only recall the many colors which we have encountered while mixing paints. The colors produced this way may in some cases seem close to those of the spectrum, but they clearly are not. They are weaker or paler, or they are darker. Another way to emphasize the vastly greater number of colors which exist beyond those seen in the spectrum is to compare the number of colors distinguishable in the spectrum to the total number of colors of any kind which can be seen and distinguished. Six colors are ordinarily seen in the spectrum of white light: violet, blue, green, yellow, orange, and red. With some further effort, particularly when viewing a bright white light through a spectroscope, the in-between colors of blue-green, green-yellow, yellow-orange, and orange-red can be seen also, bringing the total number to ten.

By contrast, the total number of colors of all kinds which can be seen is usually considered to be well over a thousand, and a trained observer, able to compare colors side by side under carefully controlled conditions, may in principle be able to distinguish thousands or millions of colors. This large range of colors which can possibly be seen – this over 1000 possibilities – presents certain problems.

The first is that most mineral collectors are not educated to the world of color. While rich and varied color surrounds us on all sides, in magazines, billboards, textiles, cosmetics, and color television, it is probably fair to say that most mineral collectors do not devote a great deal of conscious attention to color and have not the conscious purposeful understanding of color which the artist or decorator might have. Seeing before us seven different shades of red, we may lump these into one category, while the artist recognizes their clear distinctiveness. Our appreciation of color may thus be impoverished.

A second problem presented by this large number of distinguishable colors is that of color names. We are all familiar with the strange and exotic names given to lipsticks, nail polishes, and the latest colors of fashion. These names seem to convey little information to us about the nature of the color involved. To describe one color as "plum" is more specific perhaps than describing another as "shocking pink," but neither term is really very specific, very capable of conveying the same color idea to two different readers. This situation presents a problem in describing the color of fluorescence: the problem of how to describe colors so that what the writer sees is what the reader will then picture based on the color name used. More will be said about this, but a review of the nature of color is in order first.

COLOR VARIABLES

It is recognized by scientists who specialize in the study of color that any color can be completely defined and specified by a set of only three variables. These three variables are known as "hue," "saturation," and "brightness." It is useful to investigate these three attributes of color in order to obtain an interesting and important insight into how colors are formed. We may examine, for example, a group of colors: a greenish white, an intense pure green of the spectum, and the dark green color of an olive. These colors are seen to be clearly different, yet they share one attribute – greenness. Green dominates the appearance of each of these. The dominant color, in this case green, is known as a hue. However, since these colors are nonetheless different, other characteristics must be stated in order to indicate in what way these greens differ from one another. These additional characteristics are saturation brightness.

Anyone who has mixed paints has probably squeezed a tube of pigment into a white paint to create a tinted paint. The white paint dilutes the color of the pigment. This dilution can be carried out in steps. For example, suppose that the pigment is an intense green, much like the green of the spectrum. If it is mixed with a dab of white, the green will be diluted slightly. As more white is mixed in, the green will be diluted further. Finally, when sufficient white is mixed in, the greenness will be essentially invisible and the mixture will appear to be white. The different colors are identical in hue – green – but are said to differ in saturation. The pure green of the original pigment is said to be a saturated color. Another way to say this is to say that it is a pure color. As white is added, the color becomes less and less saturated. Finally, as the original green is all but completely diluted with white, it is an unsaturated color. In each step of dilution, a distinctive shade, in effect a different color, is produced.

However, the specification or description of the color is not yet complete. Colors, if they are reflective colors as in paints or inks, may differ in the amount of light they reflect; for example, some greens are darker than others, leading to a series of distinctive greens of various degrees of darkness. Colors which differ in the intensity of the light returned to the eye are said to differ in brightness. Such brightness differences can be simulated to a degree by mixing a pure green with varying amounts of black, thus decreasing reflectivity and leading to a color we perceive as darker. Unfortunately, the green pigment is also being diluted or desaturated at the same time, so that two things are happening at once. If we could view pure green through darker and darker neutral grey sunglasses, the change in perceived color due to changes only in brightness would be easier to appreciate.

For a similar reason, the example given previously to illustrate saturation is also imperfect. The problem is that white paint is almost always of higher brightness than any tube of green pigment. Thus, when the green is diluted with white, saturation decreases but brightness may increase at the same time, making the impact of saturation changes unaccompanied by brightness changes hard to discern. In place of the white, it would be preferable to use a gray paint, sufficiently dark to match the reflecting power of the original green. Mixed with progressively more gray paint, the saturation of the green would decrease steadily with no shift in brightness.

It is sometimes hard to appreciate that the darker colors owe their distinctiveness to a reduction in the light energy reaching the eye relative to a brighter color of the same hue. A black and white photograph cares little or nothing about hue and saturation, and records only the brightness of objects. Thus, in a photograph of an outdoor scene, the dark green of mature foliage will be recorded in the photograph as a dark gray, perhaps near black. The light green of new leaves will be recorded as light gray. The black and white photograph thus provides a rough gauge of the relative brightness of colors, and demonstrates how a given hue – green in this case – may vary in brightness.

COMPOSITION OF THE COLORS

All of the colors can be generated through a proper combination of a selected hue, saturation, and brightness. The appropriate starting point is the colors of the spectrum, shown in Figure 2-2 and reproduced in Figure 9-1(a). For purposes of color studies, these spectral colors are sometimes displayed around the rim of a circle to form what is called a "color wheel" as shown in Figure 9-1(b). At the center of the color wheel is white. If we start, for example, with spec-

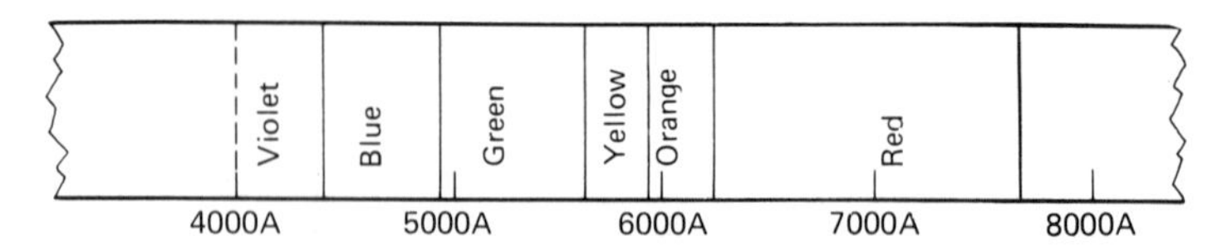

Figure 9-1a. The color spectrum of white light.

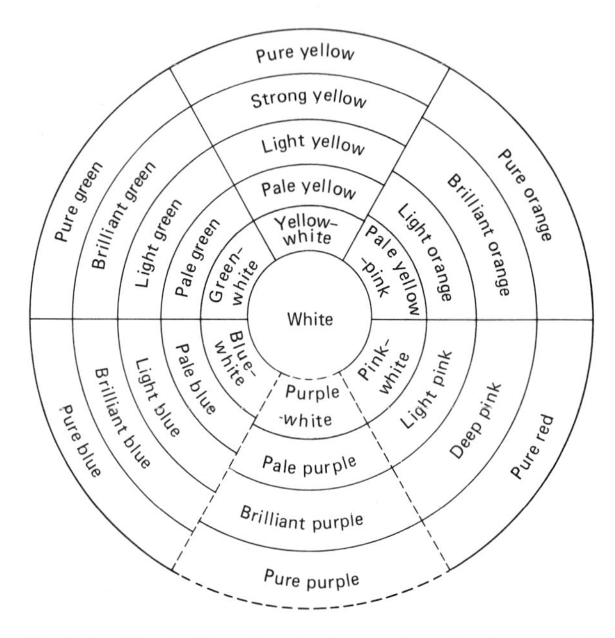

Figure 9-1b. The Color Wheel. The color wheel maps the colors in an orderly manner. The hues seen in the spectrum of white light, that is, the "pure" colors, are placed clockwise around the outer rim. Then, moving inward, the diluted or desaturated equivalents of each hue are shown. A large number of in-between colors are not shown due to lack of space. The color violet is included in the purples. The names of colors used here are intended to illustrate the idea. Various color atlases will use different color names.

tral green on the outer rim of the circle, the slightly diluted green obtained by adding a small amount of white to spectrum green is placed inward from green in the direction of white. Each step of additional dilution creates a green closer to white, until white is reached at the center. Thus, greens of decreasing saturation appear closer and closer to the center. Similarly, if we start at red and move toward the center, pink-red, pink, pinkish white, and finally white are seen. Similarly, yellow moves through paler shades until white is reached. Blue passes through shades of blue-white moving toward the center, that is, toward decreasing saturation. Along each of these steps, the hue remains the same. Thus, the color wheel provides a convenient way of mapping out the spectral hues and mapping also a large range of other colors derived by desaturating the spectral hues.

The color wheel shows a dashed line region which contains hues not seen in the spectrum. It contains a distinctive body of saturated colors, the purples, derived by mixing red and blue light in differing amounts. Depending on the relative amount of blue and red which are combined, blue-purples, magentas, and bluish reds are generated. As with the spectral colors, desaturation through mixture with white produces other colors until white is reached at the center.

If certain new colors, the purples, can be produced by mixing blue and red light, is it possible to produce still more colors by mixing together the other spectral colors? Could a color not yet mentioned be produced by mixing green and orange or blue and yellow, for example? The answer is that nothing new is produced. Rather, some other color already on the wheel will result. This color wheel thus fully illustrates in principle all of the variations of color obtainable by various hues, by combinations of hues, and by desaturation from pure color to white.

There remains one variable – brightness – which as it is changed is capable of adding many distinctive colors to those discussed above. We can imagine a second color wheel, whose colors have the same hues and same saturations as the first wheel. Only the brightness of each color has been decreased. This could be illustrated by taking a duplicate of the first wheel and illuminating it with a less intense white light. A different body of color will now be seen, particularly if directly compared to the first, more brightly illuminated wheel. These are the darker equivalents of the colors of the first wheel. For example, seen under conditions of lower and lower illumination, yellow may appear mustard or even olive-brown.

The usual way that the map of colors is completed to illustrate all combinations produced by changes in hue, saturation, and brightness, is to imagine a stack of color wheels, one above the other. Each lower wheel contains colors of lower brightness. Colors of the same hue and saturation but different brightness appear in the same vertical line, one above the other. This is illustrated in Figure 9-2. Some interesting color relationships are shown, and some interesting and important colors not yet mentioned are accounted for in this way. At the center of the brightest wheel, white is found. Straight down, white of lower illumination is seen. When compared to the first white, the second will appear gray. Further down, the appearance will be darker and darker gray until black is reached. In considering red, darker reds are encountered moving downward. A color will be reached which is identical to the color of a building brick. Thus, brick red is simply a low brightness red, identical in hue and saturation to some bright red above. Similarly, for the greens, a color is seen which is identical to the color of a green olive. Olive green is thus a low brightness green. Orange provides a particularly interesting result. If light orange of lower and lower brightness is examined, first the color tan is encountered. Then, at yet lower brightness, brown is encountered. Thus, brown is nothing more than a low brightness orange. This may seem hard to believe. One writer has

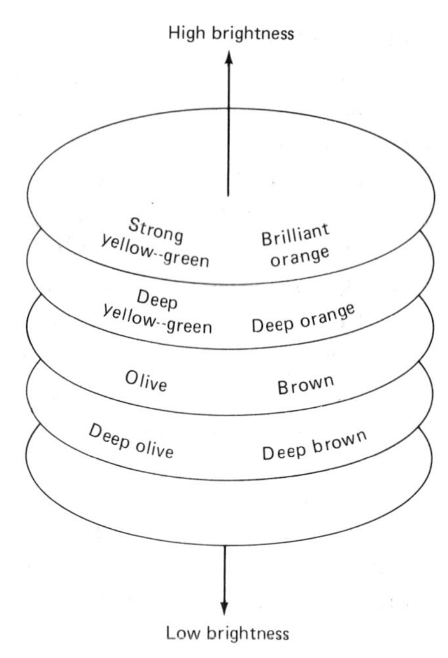

Figure 9-2. Change brightness. Brightness changes produce distinctive colors. Orange under low illumination produces brown and yellow-green produces olive. A weakly orange fluorescent mineral thus looks brown, or tan if white is present. Such darkened colors are mapped in principle by a stack of color wheels containing equivalent hues and saturations but differing in brightness.

stated, however, that the brown of a chocolate bar can be shown to be identical to orange by illuminating the chocolate with a bright light while comparing this to an orange illuminated by a weak light. The two colors will then appear to be identical.

THE BRILLIANCE OF FLUORESCENCE

All of the observable color to which the eye is sensitive are accounted for in the foregoing discussion and can in principle be located on one of the wheels of Figure 9-2. The colors of fluorescent minerals also must be included here, and they are. The colors of fluorescent minerals, in spite of the brilliance of some fluorescent colors, are not distinctive from other colors, at least in theory.

Why, then do many fluorescent minerals provide such apparently brilliant colors? Several reasons probably account for this. The colors of some mineral fluorescences are, in fact, highly saturated and pure, close to the purity of spectral colors. Measurements made on certain specimens of blue fluorescing fluorite, yellow fluorescing wernerite (scapolite), orange tremolite and wollastonite, and red halite, calcite, axinite, and ruby show that these particular specimens are nearly saturated and pure in the color of their fluorescence.* Other measurements made on the green color of fluorescing willemite and various uranium minerals indicate that, while their colors are far from saturated or pure, they are more so than most commonly seen green inks, dyes, photographic colors. As a consequence, one gains the impression that fluorescent minerals are somehow particularly colorful or brilliant. This impression is further enhanced by another favoring selection. It is usually the brightest and most intense specimens which are collected and shown regardless of shade or color. Then also, the specimens are usually viewed against the black background of a dark room, which further enhances the vividness of the fluorescence. All of these factors converge to suggest that fluorescent minerals are, as a group, somehow more brilliant, more intense, than ordinary colors. However, an examination of a broadly based collection of fluorescent minerals soon dispels the belief that all fluorescent minerals are exceptionally vivid. Many fluorescents are low in brightness, and many more are desaturated and impure in color. Many fluoresce in the dullest white or gray. Such specimens are usually quite unspectacular in fluorescent appearance, but they can be important to the collector for reasons of scientific and mineralogical interest.

A few minerals, such as some sphalerite from the Sterling mine near Franklin, New Jersey, fluoresce brick red. This can now be recognized as simply a weak or low brightness red. Given a sufficiently intense long wave ultraviolet light with which to illuminate the specimen, this sphalerite could possibly be made to fluoresce brightly enough to match a red fluorescence. Similarly, the tan or brown fluorescence seen in some norbergite or some scheelite is now recognizable as a low brightness orange. This can sometimes be seen through the use of a new, powerful, shortwave light. If this light is held sufficiently close to the specimen to force higher brightness than is usually seen, the specimen is now likely to be recognized as orange in color. Similarly, the gray or gray-white fluorescences sometimes seen are simply whites of low fluorescing power.

THE SPECTRUM OF A FLUORESCING MINERAL

The light produced by a fluorescing mineral can be viewed through a spectroscope, and by that means, the various wavelengths of which the light is composed can be seen. When this is done, it may come as a surprise to discover that the bright yellow fluorescent light of a wernerite from Canada contains, in addition to yellow light, substantial amounts of green, orange, and red. The yellow color which we see in fluorescing wernerite is the result of a merger or fusion of the colors due to these separate wavelengths. Visual perception always merges the separate colors contributed by various wavelengths of

---

*Newsome, D. and Modreski, P. 1981. The colors and spectral distributions of fluorescent minerals. *Journal of the Fluorescent Mineral Society*. Vol. 10:7–56.

light. In some cases, the result is a new color which may bear no direct or obvious resemblance to the colors that the various wavelengths of which the light is composed would separately produce.

If we examine a red fluorescing calcite or a blue fluorescing fluorite, we may wonder of what wavelengths, what components of spectral color, is this red or this blue composed. Figures 9-4 through 9-8 show the spectra of a number of fluorescent minerals. These graphs are produced by a spectrometer, which automatically measures the intensity of the light entering the instrument at each or its wavelengths. It then directs an ink plotter to trace a curve showing, by the height of the curve at each wavelength, the amount of light energy at that wavelength.

To understand how the eye deals with these wavelengths to produce the color sensations which we see, it is necessary to introduce a new map of color sensations which in some ways resembles a distorted color wheel. This is called a chromaticity diagram and is shown in Figure 9-3. As in the color wheel, the pure colors of the spectrum are wrapped around the outer rim, which in the chromaticity diagram is no longer a circle. The purples are mapped across the bottom. Then, toward the center, colors of less and less saturation are mapped until white is reached in the middle. As with the color wheel, equivalent hues or saturations of lower brightness would appear on a different map below the first, and a stack of chromaticity diagrams, much like Figure 9-2, would account for all colors. Within the diagram, various color regions are shown. As in the color wheel of Figure 9-1, these regions are only approximate, since there is simply not enough room in the figure to separately and accurately indicate and name the many shades which the eye can distinguish.

The advantage of the chromaticity diagram over the color wheel is that the chromaticity diagram can be used to tell us how the eye merges any combination of colors, whether the pure colors of the spectrum or the less saturated colors which lie within the figure.

The colors produced by color television can be used to illustrate this. The inner surface of the television picture tube face is covered with tiny phosphor dots grouped in threes, lined up in row after row over the face. One of the three dots will produce a red color when struck by an electron beam within the picture tube, a second will produce green, the third blue. Combinations of these colors in the proper ratios will produce all of the colors seen on television.

The color produced by the red phosphor dot alone is very close to a saturated color. That particular color has a location very close to the spectrum red in the chromaticity diagram. The blue phosphor produces a color slightly further removed from a spectrum blue, and the green phosphor produces a shade of green well off the spectrum greens and thus somewhat desaturated or whitened. The location of these colors is shown on the chromaticity diagram in Figure 9-3 by three points. On this diagram, straight lines have been drawn connecting these three points. Television can produce those colors, and only those colors, lying in the area enclosed by the triangle Most colors including white are inside this triangle, and television is able to reproduce these with reasonable accuracy. However, saturated purples are outside the triangle, and color television will not reproduce these but

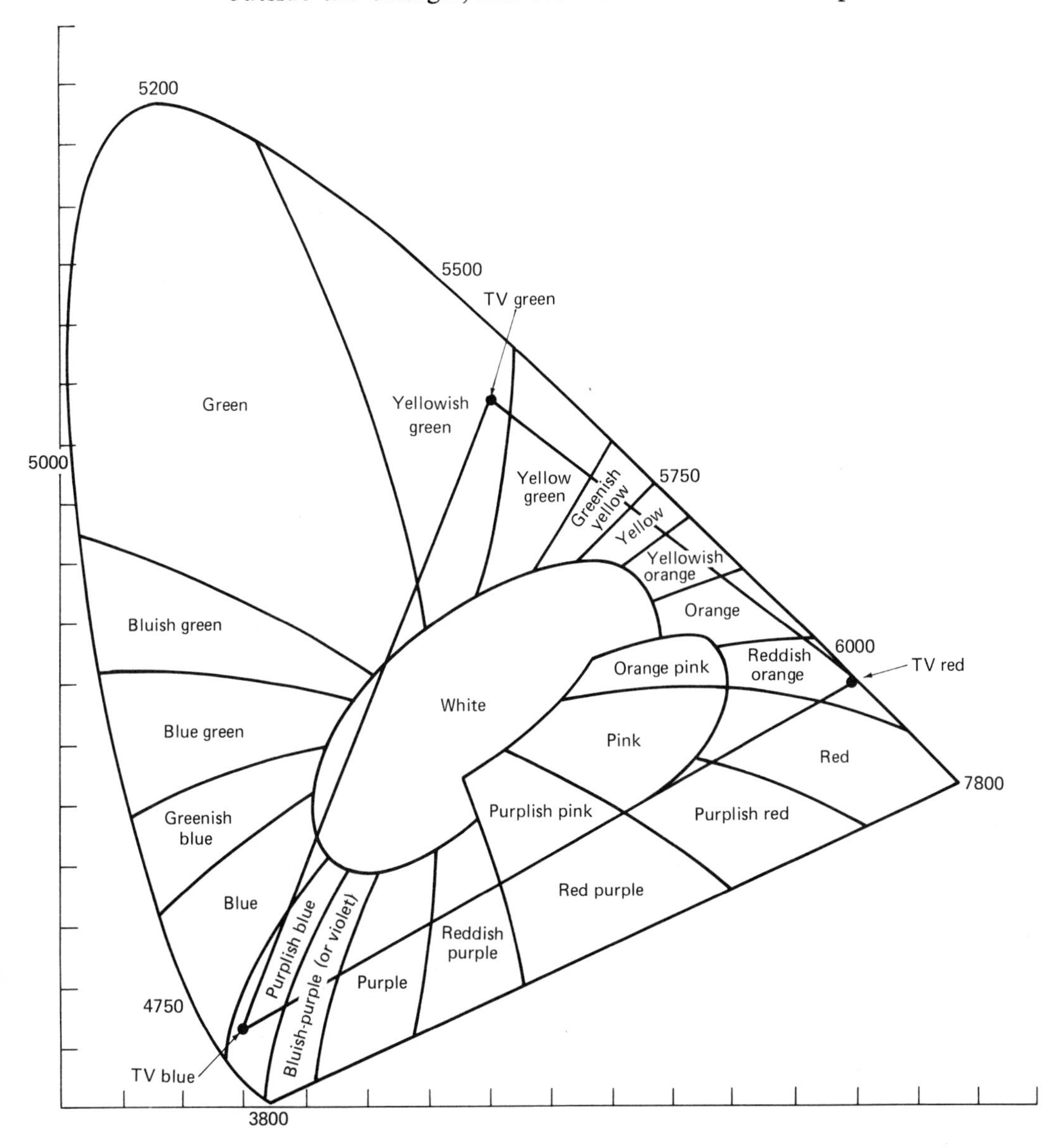

Figure 9-3. The chromaticity diagram, like the color wheel of Figure 9-1 provides a map of all colors. The pure spectrum colors are mapped along the curved portion of the outer rim. The purples appear along the straight line joining 3800A with 7800A. Less saturated colors are mapped closer to the middle until white is reached. Colors of lesser brightness would appear on other maps of this general shape, stacked one below the other much as Figure 9-2. Color regions shown on the figure are only approximate since one shade blends subtly into the next, rather than being sharply demarked as the diagram would indicate. Choice of color names for the regions is not necessarily as others would choose them. The chromaticity diagram allows measurements and judgements to be made concerning the way the eye merges or blends colors, which the color wheel does not.

will substitute a somewhat whitened version of purple. Similarly, the strongest, purest greens cannot be produced, as these are outside the triangle. In their place, some less saturated green will be shown. Extreme red and extreme blue also cannot be reproduced. In principle, television phosphors could be made which would create a larger triangle, thus allowing more colors to be reproduced, but it has been found that few people are bothered by the present color limits, providing that skin tones are reproduced accurately which they can be with the phosphors now in use.

What colors could be obtained if only two phosphor colors are used? Suppose, for example, light from only the red phosphor and the green phosphor could be used, mixed in different ratios. The answer is that any of the colors along the straight line joining red and green could be produced, and only those.

This answer is important, because it is true of any two colors, anywhere on the diagram. Pick any two colors as two specific points on the diagram. Mixing these colors will produce some color lying on the straight line joining the two points, exactly where along the line will depend on the relative brightness of the two starting colors.

Applying this rule to the chromaticity diagram leads to some interesting conclusions. For example, the red at 7800A if mixed with the blue at 3800A will produce some color lying along the line joining these points on the diagram. A line is already drawn here which shows that purples are produced by mixing red and blue, a not very surprising conclusion. However, what happens if some other colors, say blue-green at 4950A and reddish orange at 6100A, are mixed? The resulting color will lie along the line joining these colors. Depending on the brightness of the blue-green compared to the reddish orange, a desaturated blue-green or a pink could result. More surprisingly, mixing this blue-green and reddish orange could produce white. Any number of other pairs of colors can produce white. These can be found on opposite ends of any straight line drawn through the center of the white region on the diagram. Pairs of colors which produce white are called complementary colors and are discussed later. To take another example, a line could be drawn through the green at 5000A and the yellow-green at 5700A. Mixing these colors will produce some new shade of green. However, since the line does not pass close to or through the white point, the mixture can result in some desaturation (whitening) of the resulting green, but white itself will not be produced by mixing these greens.

By using the measured spectrum of fluorescent light of a mineral, it is possible to determine with high accuracy the color which will be seen when the contributing colors are merged by the eye. However, extensive calculations are required for this purpose. Nevertheless, in some cases at least, it is possible to gain some insight into the color likely to be produced by simpler means, by comparing the spectrum graph to the chromaticity diagram.

Figure 9-4(a) shows the spectrum of fluorescing ruby. This spectrum is quite narrow, covering a narrow region in the red portion of the spectrum. Since no other wavelengths are present to contribute other colors, it is to be expected that the color of the fluorescence will be a pure saturated red, and it is. By contrast, the spectrum of lead-manganese activated calcite, Figure 9-4(b), shows colors from yellow-green through yellow, orange, and red. A straight line drawn on the chromaticity diagram from yellow-green at 5600A, at one end of the spectrum of calcite fluorescence, to the other end at extreme red near 7800A will also contain all the wavelengths in between. Any two of these wavelengths, when combined in the eye, will produce a color also lying along this line; therefore, the result of fusing all these wavelengths will be a color lying somewhere along this line. Since the line is almost exactly along the pure spectrum colors, the color finally seen will be some almost pure, saturated color. Since the eye is less sensitive to red than to orange, a reddish orange color should be expected, which is what is seen in such calcites.

The spectrum of fluorescing fluorite, Figure 9-5(a), shows violet, blue, and green at lower intensity. This green extends to about 5100A. The color seen could be expected to lie perhaps close to a line connecting the bluish purple or violet end of the spectrum with the 5100A green, but closer to violet or blue. This line lies away from the spectrum color curve of the chromaticity diagram, and the result is that the color of fluorescing fluorite should be expected to be somewhat desaturated. Measurements confirm this. The color is a slightly desaturated bluish purple.

The deep sky blue fluorescent colors, under short wave, of calcites of the Terlingua type are probably familiar to most collectors. The spectrum of one such specimen is shown in Figure 9-6(a). It resembles the spectrum of fluorite in a general way except that increased amounts of green and other hues are present. The result of this change in spectrum compared to fluorite would be somewhat hard to predict without careful calculation. However, measurements indicate that the color of the fluorescence of this calcite is in the purplish blue portion of the chromaticity diagram, but halfway toward white.

This same Terlingua-type calcite fluoresces in shades of pink under long wave ultraviolet. The spectrum under long wave is shown in Figure 9-6(b). In the blue region, a large peak is evident, as in the spectrum produced under short wave, but now something new appears – a low, broad peak spread from greenish yellow through orange to red.

The color which would be produced by the new peak acting alone, in the absence of blue, is somewhat hard to judge. However, it would

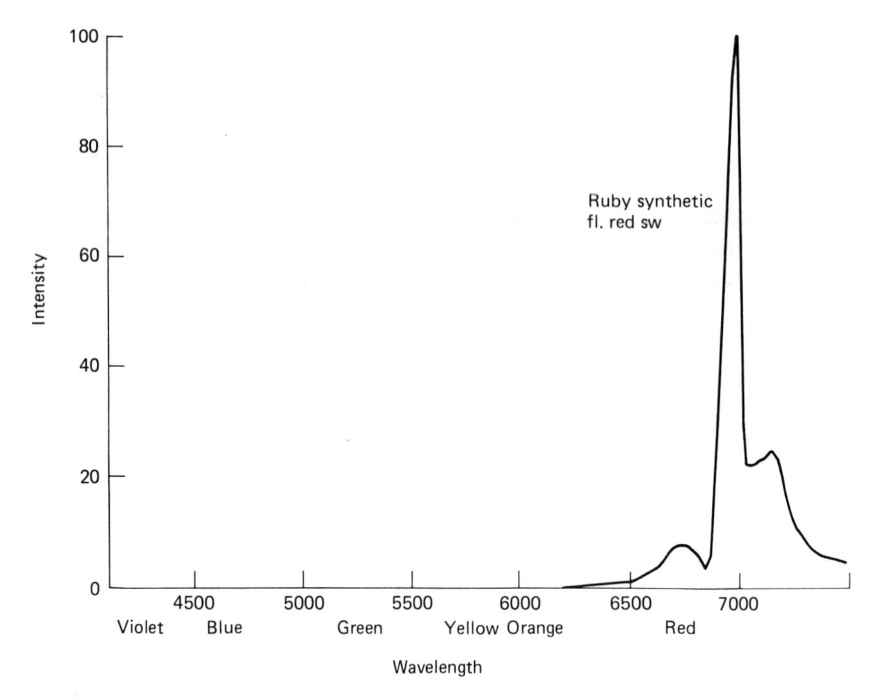

Figure 9-4a.

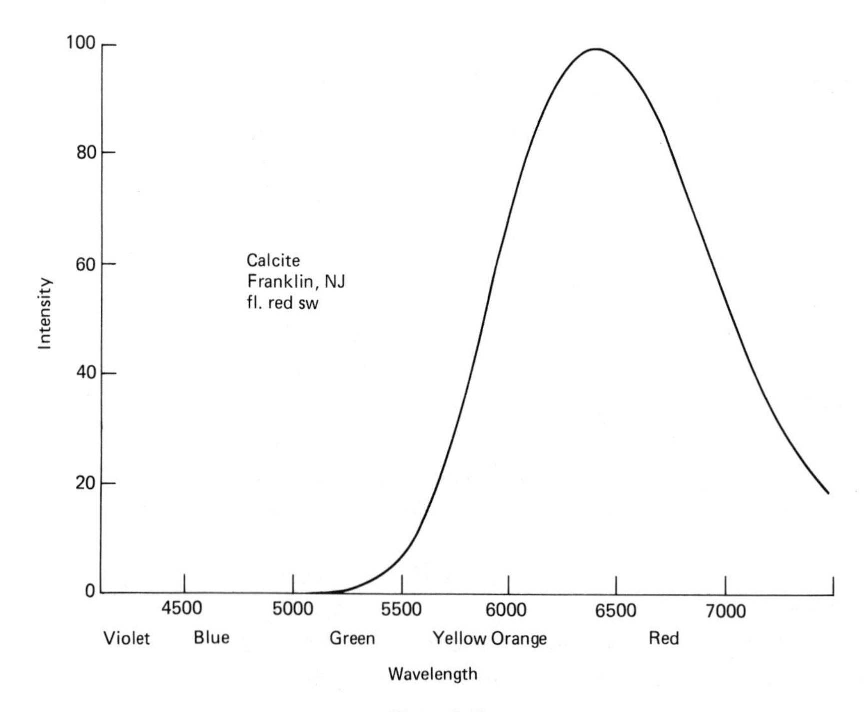

Figure 9-4b.

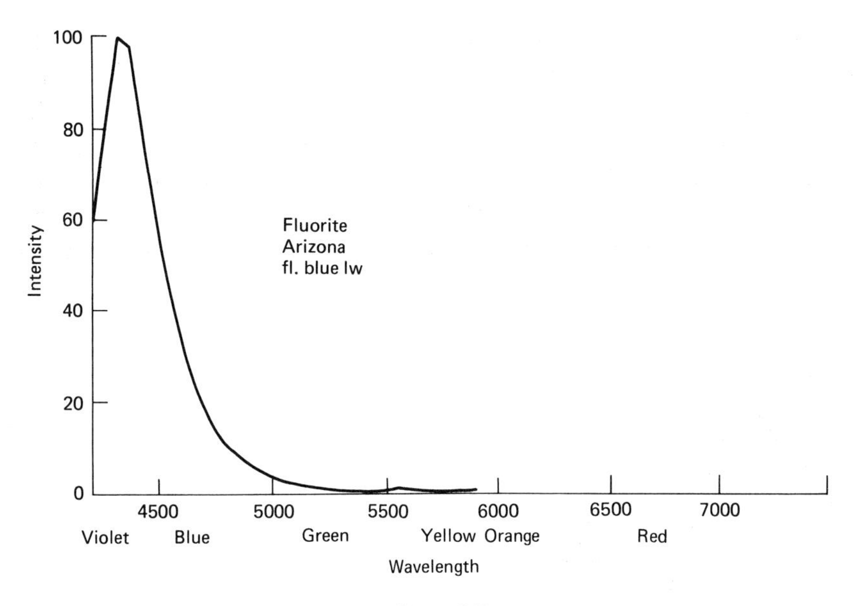

Figure 9-5a.

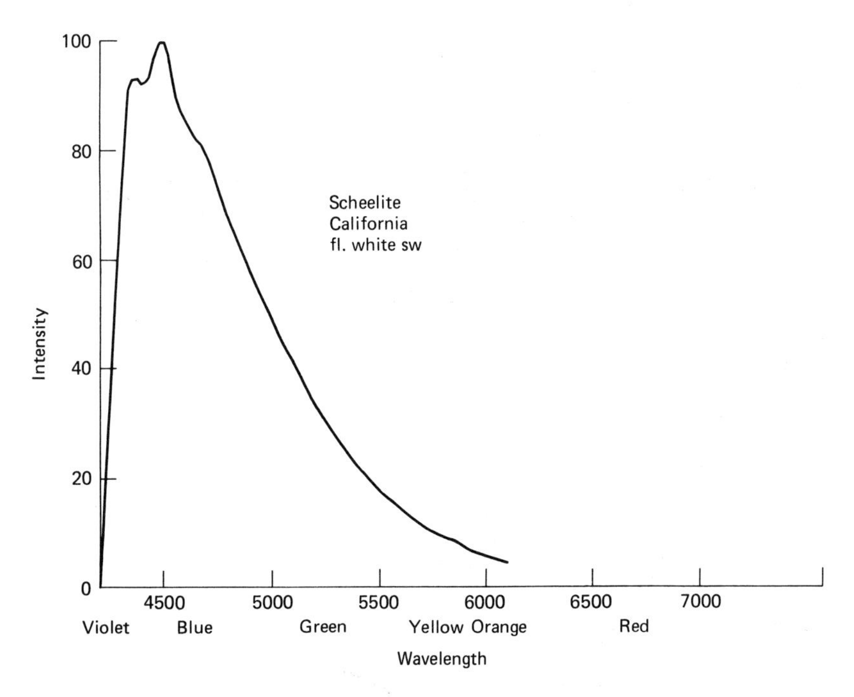

Figure 9-5b.

be expected to be some form of orange or red-orange. Careful observation of the fluorescence of Terlingua-type calcites shows that under long wave ultraviolet, there is an initial bright flash of light when the ultraviolet is first directed at the calcite. Then, the calcite settles down to its pink color. This initial flash may well be the color due to the lower peak, before the slower blue response begins to add its own blue-white light. The color of the flash varies. In one specimen examined it was orange.

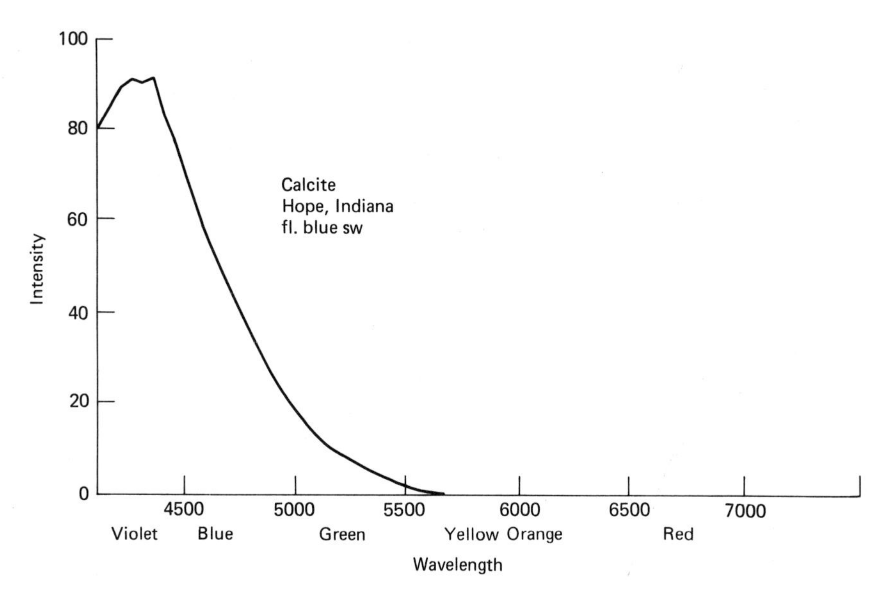

Figure 9-6a.

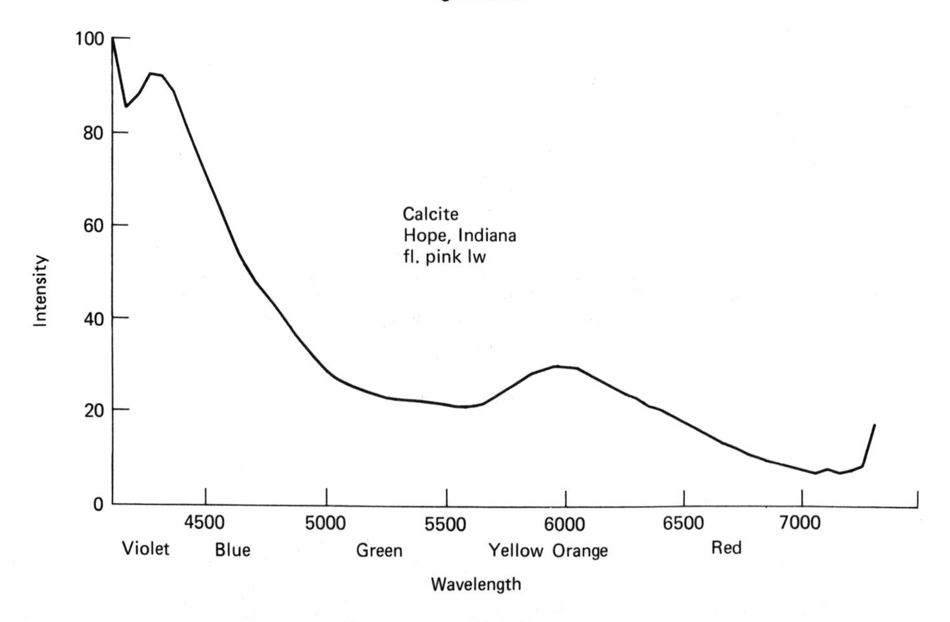

Figure 9-6b.

We can assume, then, that the long wave color of this calcite results from the fusion or merger of two colors. The first is the blue, familiar in the short wave response and apparently present in the long wave response also. The second is the new color orange, stimulated only by long wave. A straight line can now be drawn between the purplish blue region of the chromaticity diagram (where the blue color is known to be) and the orange region. The long wave color of this calcite should be expected along this line. The line passes through oranges, pinks, purplish pinks, and white. Pinks, purple-pinks, and almost white have been seen at one time or another in various Terlingua-type calcite under long wave ultraviolet.

The fluorescent spectrum of meta-autunite is more complicated than that of other minerals discussed up to this point. When this fluorescence is viewed through a spectroscope, the spectrum will reveal a number of brightly colored bands. The first band is bluish green in color; the second, a strong green with a slightly yellow tone; the third, a slightly less intense yellow-green; and the fourth, a still weaker green.

When this spectrum is plotted as a graph, the bright bands appear as a series of peaks, shown in Figure 9-7(a). The color resulting from fusion of these colors is the well-known yellowish green characteristic of so many uranium minerals. This color is found in the upper corner of the yellowish green region of the chromaticity diagram. For comparison, the spectrum of fluorescing liebigite is shown in Figure 9-7(b). It is similar to the meta-autunite spectrum, except that its bands are displaced toward the blue by several hundred angstroms with some change in the brightness of the bands. The brightest band is now green rather than a slightly yellow-green. Also, a stronger bluish green band appears. The result is as if, in the chromaticity diagram, the color has moved from the upper corner of the yellowish green down and left toward the bluish green. As a result, the yellowish coloration of the autunite fluorescence is replaced by a pure green liebigite. In crossing the chromaticity diagram toward bluish green, the color also becomes somewhat desaturated. This is why the color of fluorescent liebigite is sometimes described as whitish green; when compared side by side with autunite, the color of fluorescing liebigite appears slightly bluish by comparison, which is why it has also been described as bluish green. The spectrum and the color of schroeckingerite and andersonite are similar to that of liebigite.

QUIRKS AND DISTORTIONS IN COLOR

The human eye combines with the optic nerve and the brain to form a system of remarkable capability and adaptability. Testimony to this remarkable ability includes: the vast range of light intensity un-

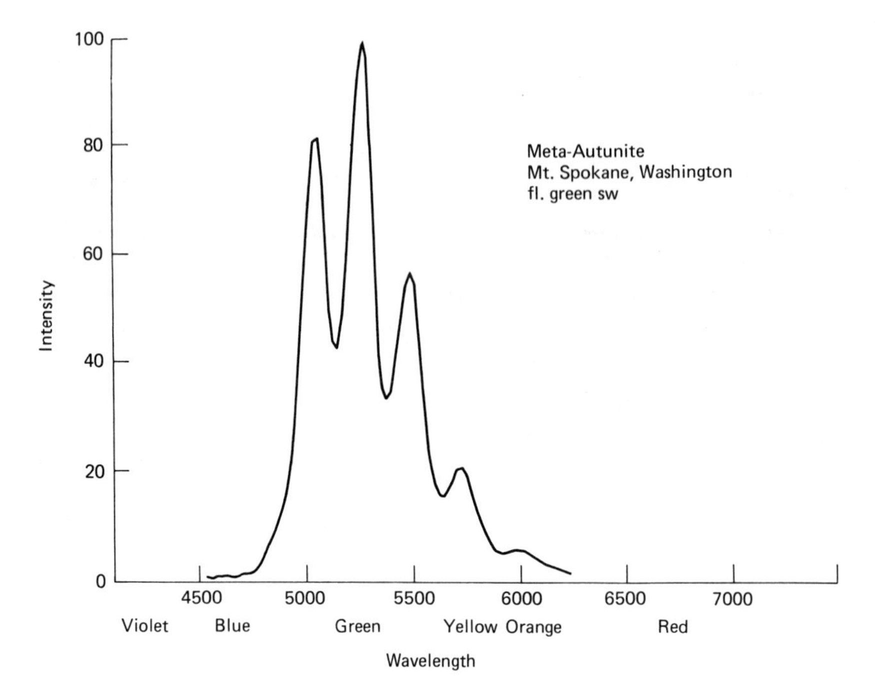

Figure 9-7a.

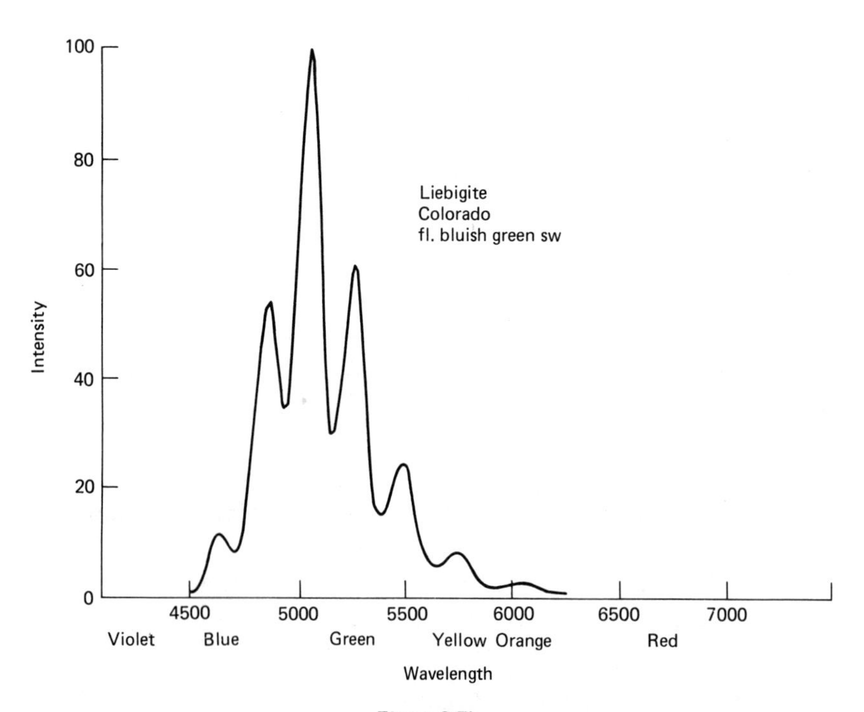

Figure 9-7b.

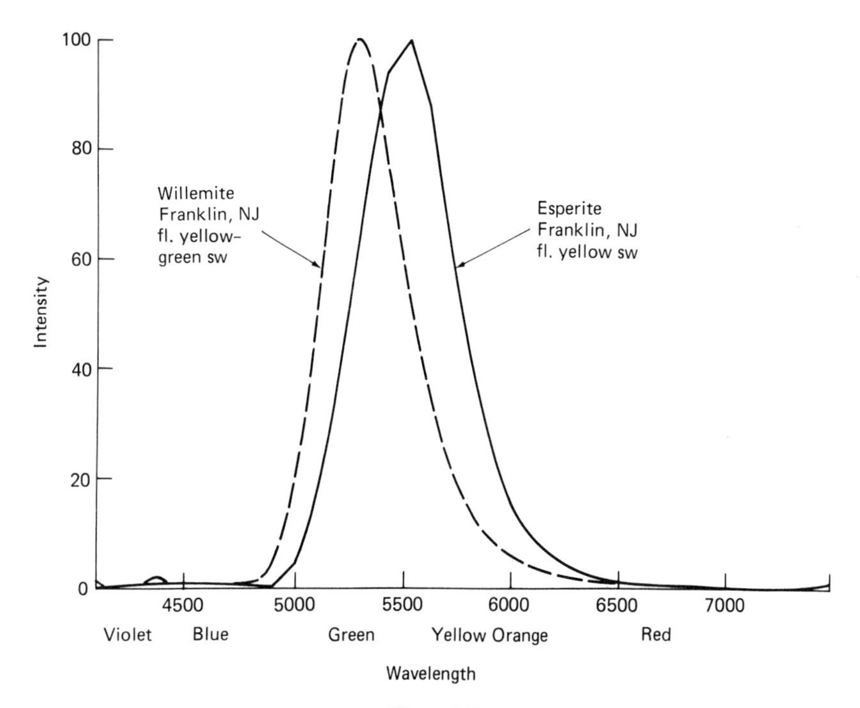

Figure 9-8a.

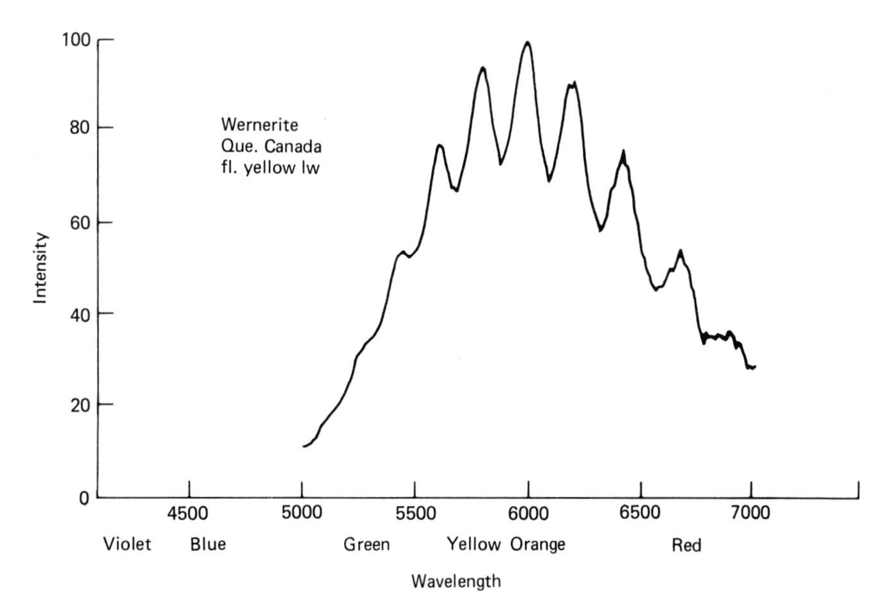

Figure 9-8b.

der which vision is possible, from the brightest sunlight to moonless starlight; the thousands, perhaps millions, of shades of color which can be distinguished; the sharpness of human vision; the ability to detect the slightest motion in the visual scene; and the recognition of forms in a jumble of differing aspects of projection and view. However, it would be surprising if these things were accomplished entirely without error, and indeed, some distortions are inevitable. These distortions are known to occur in the perception of color.

The total disappearance of color vision in night adaptation is one extreme case. When we enter a dark room, an immediate adjustment of visual sensitivity begins, and in seconds, the sensitivity of the eye may increase a hundredfold. Over the next several minutes, sensitivity increases further. After 5 to 30 minutes in darkness, the eye takes on the extreme sensitivity of dark adaptation or night vision. At that time it is possible to see reasonably well in the outdoors by starlight, but color will no longer be seen. The world will appear in shades of gray, much as in a darkened black and white television picture. If we were then suddenly to be exposed to the relatively intense illumination provided by a flashlight, or were to turn the portable ultraviolet light on a brightly fluorescent mineral, color would be seen instantly, but a new period of adaptation would be required before night vision is restored. The viewing of all but the weakest fluorescent specimens thus takes place using "daylight" vision rather than full "night" vision. At home, when we switch off the electric light and turn on the ultraviolet light to view a specimen, the same is true. The specimen is seen by the day vision color-sensitive mechanisms of the retina.

Night vision can be put to some interesting uses, however. A phosphorescent mineral can be charged and then placed by the bedside at night. Some hours later, when we awaken for a moment, the phosphorescence can often still be seen as a gray, colorless ghost of its former color and luminosity.

Other more subtle forms of color distortion can occur while examining fluorescent minerals, but it is difficult to say what the explanations are. For example, a Pugh quarry fluorite or a Franklin barite when viewed under ultraviolet light in otherwise total darkness may be butter yellow in appearance, but it will often manage to appear blue-white or green-white if just a bit of white light is superimposed, as from an electric light in a distant room.

Color distortions arising after long exposure to intense color are also known. When one examines the extensive waste dumps at Franklin, New Jersey, at night with a short wave portable, the red of calcite and the green of willemite are to be seen everywhere, but after a time – perhaps a half hour – a mysterious transformation takes place in the colors. The red of calcite is now to be found with a deep blue fluorescent mineral in place of the green. Is it scheelite, fluorite, hardystonite? The blue mineral in red calcite is now all over the dump. However, the flashlight shows the blue fluorescence to be willemite. On reexamination under the short wave light, the willemite is green once again! A shift of color had taken place.

Several kinds of distortion of color perception are known and may contribute to an explanation of these or other cases. The Purkinje effect occurs in the twilight region of visual adaptation between day and night vision. Here, blues weaken, reds disappear, and green is emphasized. In contrast, the Bezold-Brucke effect occurs under brighter conditions. In this effect, as brightness is increased, greens and green-yellows may shift toward yellow. Similarly, reddish yellow will become yellower, and blue greens will become more nearly blue. These effects, at least the green and yellow portions of them, may explain a phenomenon seen in both willemite and esperite. When a short wave light is held some distance from willemite, the color seen is green. As the light is moved closer so as to increase the brightness of fluorescence, the color becomes yellowish green and even a color close to yellow. Similarly, esperite will fluoresce greenish yellow when the short wave light is held some distance away, but it will appear an intense yellow when the light is brought closer and the intensity increases.

It should also be noted that saturation changes with brightness. Thus blues, reds, and purples, ordinarily seen as the most saturated of colors, appear to be most saturated at lower brightness but lose saturation at higher brightness. On the other hand, yellow and yellow-green gain in apparent saturation as brightness increases.

Color contrast effects can also produce certain changes or distortions in color perception. These effects have to do with complementary colors, examples of which are:

Color	Complementary Color
blue	yellow
blue-green	orange
green	red
yellow-green	purple
yellow	blue
orange	blue-green
red	green
purple	yellow-green

If a fluorescent specimen contains two minerals, one of which fluoresces in a certain color while the other fluoresces in the complement of that color, the two colors will deepen one another and will visually dramatize one another. Thus, the mixtures of red fluorescing calcite with green fluorescing willemite or yellow esperite with purple hardystonite, may owe their vividness of color to the fact that the two colors contained are complementary, or nearly so. Interestingly, if one color is nearly a complement to the other but not exactly so, the eye will make an adjustment which makes this color look closer to the correct complementary color. The blue seen after long exposure to willemite on the dump at Franklin may be due in part to a color contrast effect. The calcite in which the willemite is found fluoresces red-orange for which the complementary color is greenish blue. Sustained exposure to this red-orange may change the color seen in the willemite toward this complementary color. Why such a contrast effect should appear only after some time is not clear, however.

Another phenomenon important in the study of fluorescent minerals is the Abney effect which refers to the fact that the color of small fluorescent areas may be difficult to identify. Thus, an isolated fluorescent crystal of a few sixteenths of an inch in size, viewed at ordinary viewing distances of about 18 inches, may shift in apparent color as we examine it, while microcrystals well under a sixteenth of an inch in size may appear to be totally devoid of color. This is due to the fact that the image on the retina of such small areas covers too few of the retinal color sensors. The image size on the retina can be increased by viewing the specimen at closer range, under a magnifier, or through a microscope. In this last case, brightness will decrease, but if a sufficiently intense ultraviolet light is held close to the specimen, fluorescent color can be seen through the microscope.

Surprisingly though, in spite of these possible color distortion effects, the colors of fluorescence appear to be remarkably stable and steady. It is only under fairly unusual conditions that the shifts of color just described may be seen, and such infrequent occurrences in no way detract from the amazing beauty to be seen in most fluorescent minerals.

COLOR NAMES AND COLOR STANDARDS

National Bureau of Standards Publication 440 has attempted to consolidate the color names contained in a number of previously issued color dictionaries and atlases. The list of names is overwhelming in number, containing for example over 1500 different names for shades of red, over 2000 shades of green, and over 3000 shades of purple. Even when different names provided in different books for the same color are eliminated, the list remains overwhelming in the number of distinctive shades (hues, saturations, and brightnesses) which have been catalogued and named. This represents a formidable problem in naming the color of a mineral fluorescence, since this wealth of available names reduces rather than increases the chance that the reader will know and recognize what color is meant by any one color name.

The use of a color dictionary or color atlas could, at least in theory, provide some help. These books contain color samples or chips, and provide a name for each such chip much as in the folder of color chips to be found in the paint store. A number of famous books of this sort exist. One of the oldest was published in 1912 by Robert Ridgeway, a naturalist, who developed a dictionary of some 1100 color samples. These colors seem somehow to be unusually vivid and clean, and particularly well suited to match the vivid colors of fluorescent minerals. A dictionary by A. Maerz and M. R. Paul contains over 7000 color samples. The best known reference of this kind in this country is based on the system originated by A.H. Munsell, which has been carefully updated and expanded over the years. While this reference contains only about 1500 samples, it is structured in careful and even steps of hue, saturation, and brightness. This allows new samples to be added in between if necessary, and higher degrees of saturation to be added if superior pigments are developed.

With such standards available, it should be possible to name the color seen and to have that color understood, or is it? We view the colors in the color dictionary or atlas under some source of illumination, perhaps an electric light. As we know. various souces of illumination differ in color, the 60-watt incandescent bulb casting a yellower light than the 100-watt bulb, and both being yellower than daylight. These various colors of the illumination source change the apparent color of the chips, sometimes severely. Yet as long as we are viewing both the chip and some material whose color is to be determined, a dyed fabric for example, under the same illumination, the color of the illumination effects both chip and fabric in the same way. An accurate color match is then possible. However, when we compare a color dictionary sample illuminated by a light bulb with the color of fluorescence of a mineral, different types of illumination are involved. The color of the sample will vary with the change in light source, while the mineral will fluoresce in the nearly invariant color inherent to its makeup.

What would be required to make use of a color dictionary to describe fluorescent color is the following. First, we must both have the same dictionary with its color samples. Then, we must both own an identical and standardized light source with which to view the samples in the color dictionary in exactly the same way. Finally, we must both use the same means to accurately compare a fluorescent sample seen in a darkened room with a sample seen under a standard light, a process easily given to error. However, in principle, we can now communicate accurately concerning the color of a fluorescence. For example, I will examine a feldspar under short wave, and I will seek to match its fluorescent color to a chip in the dictionary viewed under the standard light. I will find the match to be given by the color called "dull magenta purple." You will look up this color in same dictionary while viewing the chip under the same standard light. You will now know rather exactly the color that I have seen. You may also compare some of your fluorescent minerals throught to be feldspars and find one which matches the chip, thus providing some evidence supporting the identification of your mineral.

Other, quite different methods of comparison are possible. For example, almost any color can be duplicated by combining in the proper proportions the light produced by three distinctively colored lights. Thus, a large range of different colors can be produced by selectively combining red, blue, and green light. This is exploited in color television, as described earlier.

Color measuring devices called colorimeters exploit this phenomenon and are able to utilize three standardized colored lights in such a manner as to match any visible color offered for comparison. Thus, if we both own such a colorimeter, we can communicate precisely about a fluorescent color. However, the impracticality of this process for use by mineral collectors should be obvious.

At this point, it is worth inquiring about the benefits to be gained even if either approach to standardization – the color dictionary or colorimeter – were practical. Certainly, it is highly desirable to communicate the color of fluorescent minerals in some reasonably accurate way, but is should be remembered that, in the end, nature denies us precision. The colors seen are shifty to begin with, since color perception by the eye is not fully stable and dependable as discussed earlier. Then, the actual fluorescent color of a given mineral type will vary, depending on the concentration of activator and other factors. The color of manganese activated, red fluorescing calcites can take on a number of shades: pink, red, orange-red, red with a violet tint, and so forth. There is thus little purpose in creating color standards which are more precise than the color to be seen in various samples of the mineral itself. As for a practical method of describing the colors of fluorescent minerals with precision in a way that is useful to the mineral collector, this remains an unsolved problem.

Description of Color of Fluorescent Minerals

In this book, two means are used to describe the colors of mineral fluorescence. First, a number of photographs of fluorescing minerals are included in this book, and these serve as a form of direct color description. Every effort has been made to render the reproduction of color accurate in these photographs, but there are limitations to what can be accomplished. The problem starts in the original photograph. The color films used depend on the action of three light-sensitive dyes whose colors, properly combined, can render a large range of color. However, they are not able to capture all colors accurately, and so a compromise in color rendition is immediately introduced. Further, to capture the fluorescence of a mineral on film, long time exposure is needed, and under these circumstances, the response of the film dyes changes from their intended values. While color compensating filters may be used in taking the photographs, this is only partially successful in rebalancing the dye response. Then, the book printer must take the developed color film and create three color separations. In effect, these are three photo reproductions of each original photograph, each taken through a separate color filter. These filters and this second photographic process introduce further errors in color rendition. The separations are employed to create four plates which are used to print three colors plus black. These superimposed colors provide the printed color pictures of this book. In this printing step, the inks chosen may not be accurate or the color balance may shift over the print run.

With all of the steps involved from the original fluorescence to the final production, it will be no surprise to learn that such photographs cannot be used as precise references to color. They do, however, serve to show in a surprisingly satisfactory way the approximate color, the variety, and the combinations of color to be seen in fluorescent minerals, and they do strongly suggest the beauty to be found in many fluorescent minerals which, as an advertisement might say, "must be seen to be appreciated."

In the second method of color description, used in the color description tables in Chapter 10, colors are described in an approximate way, using a series of familiar color names, descriptions, or examples. Obviously, these cannot provide high accuracy in describing a color, but this method is considered useful and serviceable for the purpose intended.

10
Color Tables for Mineral Identification

Fluorescent color can be a useful guide to mineral identification, and in some cases, it is the quickest and least expensive means. In the mines at Franklin and Ogdensburg, New Jersey, a portable battery operated short wave light provided immediate identification of the willemite ores and of the less frequent sphalerite. In scheelite properties, the short wave light is indispensable for identification of the tungsten ore which is so often indistinguishable from valueless rock in ordinary light. In the talc and tremolite mines at Balmat, New York, a portable ultraviolet light is useful in distinguishing talc strata from the interbedded strata of tremolite. There is little doubt that the portable ultraviolet light can be useful in many other mining situations once the tendencies, fluorescent habits, and trends of a mineral in each local situation are understood.

For the fluorescent mineral collector, the fluorescent color also provides an important clue to indentification of a mineral when this is otherwise unknown. To aid in mineral identification, a series of color tables are provided in this chapter. These color tables can also be useful in planning future mineral acquisitions. By reference to the tables, the more interesting, colorful, or rare fluorescents can be targeted for some future purchase, swap, or collecting expedition.

Table 10-1 deals with that strange and interesting group of fluorescents which, in certain specimens of a mineral, will fluoresce in two colors: one color under short wave and a different color under long wave ultraviolet. Little is known about the cause of this behavior. Perhaps different activators are involved, and certainly different energy levels are involved in the visible light produced. These minerals are particularly interesting and often quite beautiful, and are a prize in any fluorescent collection.

Table 10-2, the Quick Reference Color Table, is provided as an aid to mineral identification. The table includes the fluorescent minerals most likely to be found in the field or encountered when purchasing minerals. While a rare mineral will sooner or later be encountered (and totally new minerals continue to be discovered), it is certainly more likely that the as yet unidentified specimen will turn out to be a comparatively common mineral, rather than a rarity. The inclusion of such rare minerals, or of common minerals fluorescing in some rare way is avoided, since it would complicate identification and would draw attention to the less likely identification, rather than to the more likely identification.

The Quick Reference Color Table is divided into two sections, one for short wave responses, the other for long wave color responses. Each of these sections lists the fluorescent minerals under the general color categories: red, orange, yellow, green, and blue. Of course white is also included which, while not a color in the scientific sense, is certainly one in the practical sense and for the practical purpose intended here. Each color division of the table includes the various shades which share the same hue. For example, the "red" table includes pinks, the "blue" table includes blue-white, and so forth.

Now, if one knows the fluorescent color of an unidentified mineral, the first step is to refer to the tables and to examine the list of minerals known to fluoresce in that color. Quite a few minerals may be included in the list, yet in view of the hundreds of minerals known to fluoresce, the possibilities are now narrowed considerably. The next step which can be most productive in identification of the mineral is to refer to the writings which may be available describing the known minerals of that particular mine or quarry. If nothing can be found concerning this site, reference to the settings described in Chapter 5 may be useful. Comparison of the list of minerals fluorescing in that color with minerals known to occur at this or mineralogically similar sites is often sufficient to provide a useful clue to identification.

The tests ordinarily available to collectors can also be applied, including examination of mineral associations, hardness, luster, response to acid, specific gravity if a scale for this purpose is available, and crystal form if the specimen is crystallized. Each of these tests narrows the possibilities and will, hopefully, lead to an identification. Where there is some reason to believe that the specimen may be a relatively abundant mineral fluorescing in some unusual color not listed in Table 10-2, or that it may be a rarer mineral, Table 10-3 can then be used.

The Extended Color Table, Table 10-3, includes substantially all of the fluorescent minerals discussed in Part A of Chapter 6. Table 10-3 is thus a relatively complete tabulation of the fluorescent colors of minerals which are found in the United States.

In addition to the substantially larger number of minerals included, this table is divided into a total of 25 different color categories which allow better use to be made of the particular shade of color seen in the fluorescing specimen. Thus, while Table 10-2 groups a number of red related shades under the one heading "red," Table 10-3 breaks red into several subcategories which are often seen in fluorescent minerals. Now, when a specimen fluoresces deep red, for example, the few minerals which fluoresce in this shade can be considered, and those which fluoresce only pink or orange-red can be eliminated.

Fine divisions of this sort in a color table can introduce certain problems. Exactly what is meant by "deep red"? Some amount of description is provided in the tables which will help to define the color, but where there is doubt and more than one color description appears to be equally right, more than one division of the table can be used and considered when seeking an identification.

The use of ultraviolet response can now be seen as one more dimension for mineral identification which, taken together with the others, builds confidence in the conclusion. However, the reason that ultraviolet should be tried first is that, if the mineral fluoresces, it is the quickest and easiest test to apply.

The fluorescent test should not be used as a negative test, however. The fact that a red fluorescence is not seen does not prove that the specimen is not calcite, and an absence of blue fluorescence does not conclusively prove that the material in hand is not scheelite. Some mineral specimens occasionally do not fluoresce as advertised, and in other specimens, there may be a fluorescence but not of the color usually encountered.

Finally, mineral identification by means available to most collectors, even with the additional aid provided by fluorescent color response, is subject to some degree of error. Where uncertainty or doubt remains, more sophisticated means will be needed. These may involve spectrographic, microprobe, X-ray, or infrared analysis. A nearby university or industrial laboratory may be willing to undertake an analysis of an unknown specimen if it considers the problem to be an interesting one and can fit the analysis into its schedule.

Table 10-1
MINERALS WHICH
FLUORESCE IN
TWO COLORS

Table 10-1. Minerals Which Fluoresce in Two Colors (In Certain Specimens)

	Short Wave	Long Wave
Aragonite Argrigento, Sicily	white	pink
Benitoite San Benito Co., California	blue	red
Calcite Badlands, South Dakota Yuma Co., Arizona	bright green	white
Calcite Lycoming Co., Pennsylvania	blue	yellow
Calcite Terlingua, Texas, and other localities	blue	pink
Calcite Franklin, New Jersey	blue	red
Calcite Ogdensburg, New Jersey	dull red	dull yellow
Calcite Hurricane, Utah	red-orange	yellow
Eucryptite Rhodesia	red	white
Fluorite Fort Wayne, Indiana	white	violet
Fluorite Trumbull, Connecticut Franklin, New Jersey	green	blue
Fluorite (yttrian) Sussex Co., New Jersey Park Co., Colorado	yellow	blue
Fluorite Minas Gerais, Brazil	slight pink	pale blue
Hydrozincite Mapimi, Mexico	bright blue-white	peach
Manganapatite Little Switzerland, North Carolina	bright yellow	peach-pink
Microcline Bancroft, Ontario Sussex Co., New Jersey	red	dull blue
Nicholsonite Lehigh Co., Pennsylvania	white	yellow-orange
Scapolite Sparta, New Jersey	yellow	orange or pink
Scapolite Otter Lake, Quebec	deep red	white
Scheelite Huachuca Mts., Arizona	blue-white	orange or brown
Willemite Tiger, Arizona Tsumeb, South West Africa	greenish white	pale orange

Table 10-2

QUICK REFERENCE COLOR TABLES

Short and Long Wave

Quick Reference Color Table

White, Short Wave
Including gray, pink, bluish, greenish, and yellowish shades. aragonite (usually very phosphorescent) barite calcite celestite colemanite diopside fluorite hanksite hemimorphite laumontite magnesite quartz (chalcedony) scapolite scheelite selenite strontianite ulexite (usually weak) witherite wollastonite

Quick Reference Color Table

Red, Short Wave
Including deep red, pink, orange-red. anthophyllite axinite barite (usually weak) calcite corundum (much weaker than long wave) halite microcline scapolite tirodite

RED
SHORT WAVE

Quick Reference Color Table

Orange, Short Wave
Including deep orange, pink-orange, yellow-orange.
apatite cerussite chondrodite clinohedrite norbergite scapolite sodalite sphalerite tremolite wollastonite zircon

Quick Reference Color Table

Yellow, Short Wave
Including lemon yellow, yellow-orange, butter yellow, pale greenish yellow, yellow-white.
apatite aragonite barite calcite celestite cerussite chondrodite colemanite deweylite esperite fluorite hanksite laumontite norbergite phlogopite scapolite (including wernerite) scheelite selenite spodumene strontianite talc tremolite ulexite (usually weak) witherite wollastonite zircon

YELLOW
SHORT WAVE

Quick Reference Color Table

Green, Short Wave
Including yellow-green. adamite aragonite autunite calcite quartz (hyalite) quartz (opal) willemite

Quick Reference Color Table

Blue, Short Wave
Including violet-blue, purple, sky blue, blue-white.
albite benitoite calcite celestite colemanite diopside fluorite hardystonite howlite hydrozincite magnesite margarosanite microcline scheelite selenite strontianite witherite wollastonite

Quick Reference Color Table

White, Long Wave
Including gray and pink, bluish, greenish, and yellowish shades.
albite
aragonite (usually very phosphorescent)
barite
calcite
celestite
cerussite
colemanite
diopside
fluorite
hanksite
hemimorphite
laumontite
pectolite
quartz (chalcedony)
selenite
strontianite
ulexite (usually weak)
witherite
wollastonite

Quick Reference Color Table

Red, Long Wave
Including deep red, pink, orange-red.
anthophyllite axinite barite (usually weak) calcite corundum halite (usually weaker than short wave) sphalerite tirodite

RED
LONG WAVE

Quick Reference Color Table

Orange, Long Wave
Including deep orange, pink-orange, yellow-orange.
apatite cerussite clinohedrite scapolite sodalite sphalerite tremolite wollastonite zircon
Tan or Brown
chondrodite norbergite scheelite

Quick Reference Color Table

Yellow, Long Wave
Including lemon yellow, yellow-orange, butter yellow, pale greenish yellow, yellow-white.
apatite aragonite barite calcite celestite cerussite colemanite deweylite esperite fluorite hanksite laumontite pyrophyllite quartz (chalcedony) scapolite (including wernerite) scheelite selenite sphalerite spodumene strontianite talc ulexite (usually weak) witherite wollastonite zircon

Quick Reference Color Table

Green, Long Wave
Including yellow-green.
adamite
aragonite
autunite
calcite
quartz (hyalite)
quartz (opal)
willemite

Quick Reference Color Table

Blue, Long Wave
Including violet-blue, purple, sky blue, blue-white.
albite calcite celestite colemanite fluorite hardystonite howlite magnesite microcline pectolite selenite sphalerite strontianite witherite wollastonite

Table 10-3
EXTENDED COLOR TABLES
Short and Long Wave

Extended Color Table.

1. *White:* Completely colorless white. Snow white.	2. *Yellow-White:* White with a weak or very weak yellow tint.
apatite	albite
aphthitalite	analcime
calcite	aragonite
celestite	barite
dumortierite	calcite
fluorite	celestite
guerinite	colemanite
herderite	deweylite
hydroboracite	ettringite
laumontite	fluoborite
leonhardite	fluorite
nicholsonite	foshagite
pectolite	fresnoite
picropharmacolite	gaylussite
scheelite	halite
strontianite	hanksite
tincalconite	herderite
zektzerite	hydroboracite
	laumontite
	leonhardite
	melanophlogite
	meyerhofferite
	nahcolite
	plombierite
	scapolite
	scheelite
	selenite
	spodumene
	strontianite
	trona
	ulexite
	willemite
	witherite
	zektzerite

Extended Color Table.

3. *Green-White:* White with a weak or very weak green tint.

calcite
edenite
strontianite
willemite

5. *Gray-White:* Very weak white. Dirty white.

cowlesite
dypingite
fluorite
hemimorphite
leadhillite
melanophlogite
microcline (amazonite)
pirssonite
selenite
xonotlite

4. *Blue-White:* White with a weak or very weak sky blue tint.

albite
aragonite
barite
brannockite
brucite
calcite
celestite
colemanite
diopside
dumortierite
fluorite
foshagite
halite
hanksite
hydroboracite
magnesite
pectolite
picropharmacolite
pirssonite
plombierite
rosenhahnite
scheelite
selenite
spodumene
strontianite
trona
whewellite
witherite
wollastonite
zektzerite

**GREEN-WHITE,
GRAY-WHITE,
AND BLUE-WHITE
SHORT WAVE**

Extended Color Table.

6. *Deep Red:* Intense red of high saturation. Includes the dark red of microcline.	7. *Pink-Red:* Slightly whiter or less saturated than deep red. An orange or violet tint may be present in some cases.
corundum (ruby) eucryptite hyalophane microcline tirodite wickenburgite wulfenite	anthophyllite barite benstonite calcite microcline roeblingite scapolite sphalerite tirodite
8. *Orange-Red:* Red with some orange present. Fluorescence of a Franklin calcite under short wave typical.	9. *Pink:* White with a distinct red tint. Color of rose quartz.
axinite calcite calomel halite	anglesite barite benstonite calcite scapolite scheelite spodumene (kunzite)

Extended Color Table.

10. *Orange:* Color of the skin of a deeply colored California orange.

    apatite
    calcite
    clinohedrite
    jonesite
    sphalerite
    svabite
    tremolite
    wollastonite

12. *Pink-Orange:* Orange with a slightly whitened or desaturated color and with a pink or yellow-pink tint. Peach.

    apatite
    calcite
    pectolite
    prehnite
    sodalite
    sphalerite
    tilasite
    walstromite
    wickenburgite
    wollastonite

11. *Deep Orange:* Color of a burning ember of charcoal or the lit tip of a cigarette. Red-orange.

    apatite
    calomel
    sphalerite

13. *Tan:* White with a distinctly brown or pink-brown tint.

14. *Orange-Brown:* Intermediate between orange and brown. Ochre.

    apatite
    becquerelite
    phosphuranylite
    soddyite
    titanite
    wulfenite

**ORANGE,
DEEP ORANGE,
PINK-ORANGE,
TAN, AND
ORANGE-BROWN
SHORT WAVE**

Extended Color Table.

15. *Yellow:* Color of a lemon skin.	16. *Yellow-Orange:* Or orange-yellow. Yellow with a distinctly orange tint or orange with a substantial yellow content.
apatite	apatite
barite	calcite
calcite	calcite (strontio-)
calcite (strontio-)	cerussite
cerussite	chondrodite
chondrodite	curtisite
colemanite	fluorite (yttrian)
esperite	idrialite
fluoborite	microcline
fluorite	norbergite
leadhillite	scapolite
matlockite	scheelite
meyerhofferite	shortite
miserite	tveitite
norbergite	willemite
phlogopite	wollastonite
powellite	zircon
priceite	
scheelite	
shortite	
strontianite	
tourmaline (uvite)	
wavellite	
willemite	
wollastonite	

Extended Color Table.

17. *Butter Yellow:* Light yellow of the color of butter or the core of a banana.	18. *Pale Greenish Yellow:* Similar to butter yellow or lemon yellow but with a detectable green tint.
barite	curtisite
calcite	datolite
cerussite	edenite
colemanite	esperite
deweylite	fluorite
fluoborite	idrialite
fluorite	strontianite
fresnoite	talc
meyerhofferite	
phlogopite	
scapolite	
shortite	
spodumene	
tremolite	
trona	
wavellite	
willemite	
witherite	
wollastonite	

BUTTER YELLOW AND PALE GREENISH YELLOW SHORT WAVE

Extended Color Table.

19. *Yellow-Green:* Intense green with a definite yellow tint. Fluorescence of hyalite or autunite typical.	20. *Green:* Green without trace yellow. May contain blue tint.
abernathyite	andersonite
adamite	apophyllite
analcime	chabazite
andersonite	fluorite (chlorophane)
aragonite	liebigite
autunite	schroeckingerite
bayleyite	variscite
calcite	willemite
curtisite	
datolite	
gyrolite	
heinrichite	
idrialite	
liebigite	
meta-uranocircite	
metazeunerite	
monohydrocalcite	
novacekite	
quartz (hyalite)	
quartz (opal)	
rutherfordite	
schoepite	
schroeckingerite	
swartzite	
uranopilite	
uranospinite	
willemite	
zippeite	

Extended Color Table.

21. *Deep Violet-Blue:* Dark blue with a definite violet tint. Includes dark purple with a violet tint.

    hardystonite

22. *Violet-Blue:* Medium blue with a definite violet tint. Fluorescence of fluorite under short wave typical.

    feldspar (barium)
    fluorite
    microcline

23. *Deep Sky Blue:* Color of the sky at its bluest.

    allophane
    benitoite
    calcite
    hyalophane
    hydrozincite
    karpatite
    margarosanite
    pabstite
    scheelite
    tourmaline (elbaite)

24. *Sky Blue:* Color of the blue sky. White with a distinct light blue tint.

    albite
    brucite
    calcite
    diopside
    dumortierite
    howlite
    hydrozincite
    margarosanite
    microcline
    scheelite
    wavellite
    witherite

25. *Lavender:* Light violet-gray or light purple-gray.

    barite
    barylite
    howlite
    margarosanite
    scapolite

DEEP VIOLET-BLUE,
VIOLET-BLUE,
DEEP SKY BLUE,
SKY BLUE,
LAVENDER
SHORT WAVE

Extended Color Table.

26. *White:* Completely colorless white. Snow white.	27. *Yellow-White:* White with a weak or very weak yellow tint.
barite	albite
calcite	aragonite
celestite	barite
colemanite	benstonite
fluorite	calcite
forsterite	celestite
gowerite	cerussite
guerinite	colemanite
hedyphane	creedite
hydroboracite	datolite
laumontite	diopside
picropharmacolite	ettringite
smithsonite	fluellite
strontianite	fluorite
tincalconite	halite
	heulandite
	hydroboracite
	mesolite
	meyerhofferite
	nahcolite
	plombierite
	quartz (chalcedony)
	selenite
	spodumene
	strontianite
	sulfohalite
	szaibelyite
	trona
	ulexite
	witherite
	wollastonite

Extended Color Table.

28. *Green-White:* White with a weak or very weak green tint.

aragonite
calcite

29. *Blue-White:* White with a weak or very weak sky blue tint.

albite
aragonite
benstonite
brucite
calcite
celestite
colemanite
diopside
dypingite
eucryptite
gowerite
halite
hydroboracite
magnesite
mesolite
meyerhofferite
natrolite
pectolite
pirssonite
selenite
strontianite
thomsonite
trona
wavellite
whewellite
witherite
wollastonite

30. *Gray-White:* Very weak white. Dirty white.

aragonite
barite
cerussite
cowlesite
datolite
eucryptite
hemimorphite
pirssonite
smithsonite
stilbite
thaumasite
thomsonite
xonotlite

GREEN-WHITE, BLUE-WHITE, AND GRAY-WHITE LONG WAVE

Extended Color Table.

31. *Deep Red:* Intense red of high saturation. Includes dark red of microcline. anthophyllite corundum (ruby) tirodite wulfenite	32. *Pink-Red:* Slightly whiter or less saturated than deep red. An orange or violet tint may be present in some cases. anthophyllite barite benitoite benstonite bustamite calcite corundum (sapphire) halite hedyphane scapolite sphalerite spodumene
33. *Orange-Red:* Red with some orange present. Fluorescence of a Franklin calcite under short wave typical. axinite calcite calomel	34. *Pink:* White with a distinct red tint. Color of rose quartz. anglesite barite benstonite calcite gaylussite scapolite

Extended Color Table.

35. *Orange:* Color of the skin of a deeply colored California orange.

    barite
    calcite
    clinohedrite
    margarosanite
    sodalite
    sphalerite
    tremolite
    wollastonite

36. *Deep Orange:* Color of a burning ember of charcoal or the lit tip of a cigarette. Red-orange.

    calomel
    sodalite
    sphalerite
    wollastonite

37. *Pink-Orange:* Orange with a slightly whitened or desaturated color and with a pink or yellow-pink tint. Peach.

    clinohedrite
    gaylussite
    pectolite
    scapolite
    sodalite
    sphalerite
    walstromite
    wollastonite

38. *Tan:* White with a distinctly brown or pink-brown tint.

    barite
    calcite
    celestite
    scheelite

39. *Orange-Brown:* Intermediate between orange and brown.

    apatite
    becquerelite
    boltwoodite
    phosphuranylite
    scheelite
    soddyite
    willemite
    wulfenite

**ORANGE,
DEEP ORANGE,
PINK-ORANGE,
TAN, AND
ORANGE-BROWN
LONG WAVE**

Extended Color Table.

40. *Yellow:* Color of a lemon skin.	41. *Yellow-Orange:* Or orange-yellow. Yellow with a distinctly orange tint or orange with a substantial yellow content.
barite	anglesite
calcite	apatite
calcite (strontio-)	beryl
cerussite	calcite
colemanite	calcite (strontio-)
curtisite	cerussite
esperite	curtisite
fluorite	idrialite
idrialite	microcline
leadhillite	nicholsonite
matlockite	scapolite
sabugalite	shortite
shortite	smithsonite
sklodowskite	sphalerite
strontianite	szaibelyite
wavellite	tremolite
zincite	trona
	willemite
	wollastonite
	zircon

Extended Color Table.

42. *Butter Yellow:* Light yellow of the color of butter or the core of a banana.	43. *Pale Greenish Yellow:* Similar to butter yellow or lemon yellow but with a detectable green tint.
analcime	aphthitalite
apophyllite	calcite
aragonite	curtisite
barite	esperite
cahnite	fluorite
calcite	idrialite
celestite	zincite
cerussite	
colemanite	
deweylite	
diopside	
esperite	
fluorite	
foshagite	
hanksite	
hemimorphite	
melanophlogite	
nahcolite	
plombierite	
powellite	
pyrophyllite	
quartz	
quartz (chalcedony)	
rosenhahnite	
spodumene	
talc	
tremolite	
trona	
uralolite	
wavellite	
witherite	
wollastonite	

BUTTER YELLOW AND PALE GREENISH YELLOW LONG WAVE

Extended Color Table.

44. *Yellow-Green:* Intense green with a definite yellow tint. Fluorescence of hyalite or autunite typical.	45. *Green:* Green without trace yellow. May contain a blue tint.
abernathyite	andersonite
adamite	apophyllite
analcime	chabazite
andersonite	fluorite (chlorophane)
aphthitalite	liebigite
apophyllite	schroeckingerite
aragonite	variscite
autunite	willemite
bayleyite	
calcite	
chabazite	
gyrolite	
heinrichite	
liebigite	
meta-uranocircite	
metazeunerite	
monohydrocalcite	
novacekite	
quartz (opal)	
quartz (hyalite)	
rutherfordite	
schoepite	
schroeckingerite	
swartzite	
uranophane	
uranopilite	
uranospinite	
willemite	
zippeite	

Extended Color Table.

46. *Deep Violet-Blue:* Dark blue with a definite violet tint. Includes dark purple with a violet tint.

    hardystonite

47. *Violet-Blue:* Medium blue with a definite violet tint. Fluorescence of fluorite under short wave typical.

    barite
    calcite
    celsian
    fluorite
    microcline
    sphalerite

48. *Deep Sky Blue:* Color of the sky at its bluest.

    allophane
    karpatite
    margarite
    microcline
    scapolite
    sphalerite

49. *Sky Blue:* Color of the blue sky. White with a distinctly light blue tint.

    brucite
    dypingite
    howlite
    microcline
    pectolite
    wavellite

50. *Lavender:* Light violet-gray or light purple-gray.

    apatite
    barite
    hedyphane
    hydrozincite

**DEEP VIOLET-BLUE,
VIOLET-BLUE,
DEEP SKY BLUE,
SKY BLUE, AND
LAVENDER
LONG WAVE**

Bibliography

GENERAL REFERENCES ON MINERAL FLUORESCENCE

Barnes, D. 1958. *Infrared Luminescence of Minerals.* Washington: Government Printing Office.

Dake, H. C. 1953. *The Uranium and Fluorescent Minerals.* Portland: Mineralogist Publishing Co.

Gleason, S. 1960. *Ultraviolet Guide to Minerals.* Princeton, N. J.: Van Nostrand.

Lieber, W. *Leuchtende Kristalle.* Wiesloch, Germany: Vetter.

Warren, T. 1969. *Minerals that Fluoresce with Mineralight Lamps.* San Gabriel, Calif.: Ultra Violet Products, Inc.

Raytech. 1965. *The Story of Fluorescence.* Stafford Springs, Conn. Raytech Industries.

CHAPTER 1

Gunnell, E. M. 1939. Historical notes – mineral luminescence. *The Mineralogist* **7** (3): 75–76.

Gunnell, E. M. 1939. Bibliography of mineral luminescence. *The Mineralogist* **7** (3): 81–83, 135.

Harvey, E. N. 1957. *A History of Luminescence.* Philadelphia: American Philosophical Society.

Kunz, G. F. and Baskerville, C. 1903. The action of radium, roentgen rays and the untra-violet light on minerals and gems. *Science* **18** (468): 669–783.

CHAPTERS 2 AND 3

Andrews, W. S. 1916. Apparatus for producing ultra-violet radiation. *General Electric Review* **19**: 317–319.

Boast, W. B. 1953. *Illumination Engineering.* New York: McGraw-Hill.

Gunnell, E. M. 1935. New Jersey willemites show spectacular fluorescence. *The Mineralogist* **3** (1): 9–10, 22.

Koller, L. R. 1953. *Ultraviolet Radiation.* New York: John Wiley & Sons.

Palache, C. 1928. The phosphorescence and fluorescence of Franklin minerals. *The American Mineralogist* **13** (7): 330–333.

Roth, L. H. 1959. Solarization ultraviolet filters. *The Mineralogist* 151–152.

Roth, L. H. 1960. Getting the most out of your ultraviolet lamp. *Earth Science* **13** (4): 141–143.

Roth, L. H. 1961. Small ultraviolet lamps and portability. *Earth Science* **14** (3): 119–121, 131.

Spencer, L. J. 1929. Fluorescence of minerals in ultra-violet rays. *Journal of the Mineralogical Society of America.* **14** (1): 33–37.

Summer, W. 1962. *Ultra-Violet and Infra-Red Engineering.* New York: Interscience Publishers.

Product Catalogs

Color Filter Glasses. Corning, N.Y.: Corning Glass Works.
High Intensity Discharge Lamps. General Electric.
Hoya Color Filter Glass. Tokyo, Japan: Hoya Corp.
Lamp Bulletin. General Electric.
Lamp Catalog 9200 General Electric.
Schott Color Filter Class. Duryea, Pa.: Schott Optical Glass, Inc.
Ultraviolet Sources, Meters, Components for Research and Industry. San Gabriel, Calif.: Ultra-Violet Products.
1982 Catalog. Stafford Springs, Conn.: Raytech Industries.

CHAPTERS 5 AND 6

Alfors, J. T., Stinton, M. C., and Matthews, R. A. 1965. Seven new barium minerals from eastern Fresno County, California. *American Minerologist* **50**: 314–340.

Anthony, J. W., Williams, S. A., and Bideaux, R. A. 1977. *Mineralogy of Arizona.* Tucson: University of Arizona Press.

Aristarain, L. F. 1972. Boron minerals and deposits. *The Mineralogical Record* **3**: 213–219.

Bastin, E. S. 1931. The fluorspar deposits of Hardin and Pope counties, Illinois. *State Geological Bulletin* **58**: 1–116.

Bateman, A. M. 1950. *Economic Mineral Deposits.* New York: John Wiley & Sons.

Ben-Avraham, Z. 1981. The movement of continents. *American Scientist* **69**: 291–299.

Bostwick, R. 1977. The fluorescent minerals of Franklin and Sterling Hill, New Jersey. *Journal of the Fluorescent Mineral Society* **6** (1): 7–40.

Breemen, O. van and Bluck, B. J. 1981. Episodic granite plutonism in the Scottish Caledonides. *Nature* **291**: 113–117.

Brobst, D. A 1962. Geology of the Sprude Pine district Avery, Mitchell, and Yancey counties North Carolina. *Geological Survey Bulletin* **1122-A**: 1–26.

Carpenter, A. B. 1967. Mineralolgy and petrology of the system CaO-MgO-CO_2-H_2O at Crestmore, California. *The American Mineralogist* **52**: 1341–1361.

Crook, W. W., III. 1978. Tveitite from the Barringer Hill district, Texas. *Mineralogical Record* **9**: 387.

Currier, L. W. and Hubbert, M. K.1944. Geological and geophysical survey of fluorspar areas in Hardin County Illinois. *Geological Survey Bulletin* **942**: 1–147.

Devito, F., Parcel, R. T., Jr., and Jefferson, G. T. 1971. Contact metamorphic minerals at Crestmore quarry, Riverside, California. Field Trip No. 8: 94–124.

Dunn, P.J. 1975. Rosenhahnite, a second occurrence. *Mineralogical Record* **6**: 300-301.

Dunn, P. J. 1977. Fluellite from North Carolina. *Mineralogical Record* **8**: 392–393.

Dunn, P. J. 1979. Contributions to the mineralogy of Franklin and Sterling Hill, New Jersey. *Mineralogical Record* **10**: 160–165.

Elberty, W. and Lessing, P. 1971. Economic geology of International talc and Benson Iron mines. *New York Geological Association Guide Book* Trip D.

Engel, A. E. 1956. Apropos the Grenville. *The Royal Society of Canada Special* **1**: 74–95.

Evans, H. F. 1980. Historical sketches of copper and lead mining in Montgomery County, Pennsylvania. *Friends of Mineralogy, Pennsylvania Chapter, Inc.* **2**: 6–34.

Evans, J. R., Taylor, G. C., and Rapp, J. S. 1976. Mines and mineral deposits in Death Valley National Monument, California. *California Division of Mines and Geology* **125**: 7–60.

Fersman, A. E. 1926. Minerals of the Kola Peninsula. *The American Mineralogist* **11**: 289–298.

Fleisher, M. 1980. Glossary of Mineral Species, Tuscon, Arizona: *Mineralogical Record.*

Flink, G. 1926. Langban and its minerals. *The American Mineralogist* **11**: 195–199.

Frondel, C. and Baum, J. L. 1974. Structure and mineralogy of the Franklin zinc-iron-manganese deposit, New Jersey. Reprinted from *Economic Geology* **69**: 157–180.

Guest, J. E. and Duncan, A. M. 1981. Internal plumbing of Mount Etna. *Nature* **290**: 584–586.

Gunnell, E. M. 1933. The photo-luminescence of Illinois fluorite and certain zinc minerals and associated species from the Joplin, Missouri, district. *Journal of the Mineralogical Society of America* 68–73.

Heinrich, E. W. 1958. *Mineralogy and Geology of Radioactive Raw Materials.* New York: McGraw-Hill.

Hilpert, L. S. 1969. Uranium resources of northwestern New Mexico. *Geological Survey Professional Paper* **603**: 1–166.

Holmes, R. W. and Kennedy, M. B. 1983. *Mines and Minerals of The Great American Rift.* New York: Van Nostrand Reinhold.

Jahns, R. H. and Wright, L. A. 1951. Gem-and lithium-bearing pegmatites of the Pala district, San Diego County, California. *Paper of the State of California Department of Natural Resources* **7-A**: 1–72.

Jones, R. W., Jr. 1981. *Franklin, Fluorescent Mineral Capital of the World.* Spex Industries Inc.

Jones, R. W., Jr. 1970. *Nature's Hidden Rainbows.* San Gabriel, Calif.: Ultra-Violet Products, Inc.

Kerr, P. F. 1958. Tungsten mineralization in the United States. *Geological Society of America* **15**: 1–23.

Kerr, P. F., Brophy, G. P., Dahl, H. M., Green, J., and Woolard, L. E. 1957. Marysvale, Utah, uranium area. *The Geological Society of America Special Paper* **64**: 1–77.

Lingren, W. 1933. *Mineral Deposits.* New York: McGraw-Hill.

Miller, W. V. 1975. Fluorescent minerals of the Balmat-Fowler area. *Journal of the Fluorescent Mineral Society* **4** (1): 4–9.

Miser, H. D. and Glass, J. J. 1941. Fluorescent sodalite and hackmanite from Magnet Cove, Arkansas. *American Mineralogist* 437 – 444.

Moench, R. H. and Schlee, J. S. 1967. Geology and uranium deposits of the Laguna district, New Mexico. *Geological Survey Professional Paper* **519**: 1–100.

Morgan, V. and Erd, R. C. 1969. Minerals of the Kramer borate district, California. *Mineral Information Service* 143–152.

Murdoch, J. 1961. Crestmore, past and present. *American Mineralogist* **46**: 249–256.

Murdoch, J. and Webb, R. W. 1966. *Minerals of California.* California Division of Mines and Geology.

Newhouse, W. H. (Ed.). 1942. *Ore Deposits as Related to Structural Features.* Princeton, N.J.: Princeton University Press.

Nininger, R. D. 1956. *Minerals for Atomic Energy.* Princeton, N.J.: Van Nostrand.

Nissen, A. L. Fluorescent hydrogrossular from Nordland, Norway. *American Mineralogist* **57**: 1535–1540.

Palache, C. 1929. A comparison of the ore deposits of Langban, Sweden, with those of Franklin, New Jersey. *The American Mineralogist* **14**: 43-47.

Palache, C. 1935. The minerals of Franklin and Sterling Hill Sussex County, New Jersey. *U. S. Geological Survey Paper 180.*

Park, C. F., Jr. and MacDiarmid, R. A. 1964. *Ore Deposits.* San Francisco, Calif.: Freeman.

Parker, F. J. and Peters, T. A. 1978. Tilasite from the Sterling Hill Mine, Ogdensburg, New Jersey. *Mineralogical Record* **9**: 385ff.

Pemberton, H. E. 1978. Hydroboracite from the Furnace Creek formation: a historical note. *Mineralogical Record* **9**: 379.

Pemberton, H. E. 1983. *Minerals of California.* New York: Van Nostrand Reinhold.

Petersen, E. 1979. *Fluorescent Minerals from Balmat-Edwards.* Department of Earth Science, Dartmouth College, Hanover, N.H.

Ramberg, H. 1958. *The Origin of Metamorphic and Metasomatic Rocks.* Chicago: University of Chicago Press.

Roberts, W. L., Rapp, G. R., Jr., and Weber, J. 1974. *Encyclopedia of Minerals.* New York: Van Nostrand Reinhold.

Rossman, G. R. and Squires, R. L. 1974. The occurrence of alstonite at Cave in Rock, Illinois. *Mineralogical Record* **5**: 266–269.

Santos, E. S. 1978. Stratigraphy of the Morrison formation and structure of the Ambrosia Lake District, New Mexico. *Geological Survey Bulletin* **1272-E**: 1–30.

Sawkins, F. J. 1972. Sulfide deposits in relation to plate tectonics. *Journal of Geology* **80**: 377–397.

Segeler, C. G. 1961. The first occurrence of manganoan cummingtonite, tirodite. *American Mineralogist* **46**: 637ff.

Sillitoe, R. H. 1972. A plate tectonic model for the origin of porphyry copper deposits. *Economic Geology* **67**:184–197.

Sillitoe, R. H. 1972. Relation of metal provinces in western America to subduction of oceanic lithosphere. *Geological Survey of America Bulletin* **83**:813–817.

Smith, G. I. 1972. Subsurface stratigraphy and geochemistry of late quarternary evaporites, Searles Lake, California. *Geological Survey Professional Paper* **1043**:1–130.

Speer, A. 1976. Strontianite. *Mineralogical Record* **7**:69–71.

Stanton, R. L. 1972. *Ore Petrology.* New York: McGraw-Hill.

Sutulov, A. 1974. *Copper Porphyries.* Salt Lake City, Utah: University of Utah Printing Services.

Titley, S. R. and Hicks, C. L. 1966. *Geology of the Porphyry Copper Deposits Southwestern North America.* Tucson: University of Arizona Press.

Titley, S. R. 1981. Porphyry copper. *American Scientist* **69**:632–638.

Trace, R. D. 1962. Geology and fluorspar deposits of the Levias-Keystone and Dike-Eaton areas, Crittenden County, Kentucky. *Geological Survey Bulletin* **1122**-E:13–17.

Turner, F. J. and Verhoohen, J. 1960. *Igneous and Metamorphic Petrology.* New York: McGraw-Hill.

de Waard, D. 1964. Notes on the geology of the south-central Adirondack Highlands. *New York Geological Association Guide Book* 3–14.

White, J.S., Arem, J.E., Nelen, J.A., Leavens, P.B., and Thomssen, R.W. Brannockite, a new tin mineral. *Mineralogical Record* **4**(2):73–78.

Widmer, K. 1964. *The Geology and Geography of New Jersey.* Princeton, Van Nostrand.

Winkler, H. G. F. 1974. *Petrogenesis of Metamorphic Rocks.* New York: Springer-Verlag.

Wise, W. S. and Pabst, A. 1977. Jonesite, a new mineral from the benitoite gem mine, San Benito County, California. *Mineralogical Record* 8:453–456.

Yates, R. G. and Thompson, G. A. 1959. Geology and quicksilver deposits of the Terlingua district, Texas. *Geological Survey Professional Paper* **312**:66.

CHAPTERS 7 AND 8

Avouris, P. and Morgan, T. N. 1981. A tunneling model for the decay of luminescence in inorganic phosphors: the Case of Zn_2SiO_4Mn. *The Journal of Chemical Physics* **74**:4347–4355.

Banks, E. and Schwartz, R. W. 1969. Phosphorescence mechanism in $CdF_2:Eu^3$. *The Journal of Chemical Physics* **51**:1956–1959.

Blasse, G. 1980. The luminescence of closed-shell transition-metal complexes: new developments. *Structure and Bonding* **42**:1–41.

Blasse, G. and Bril, A. 1967. Investigations on Bi^{3+} -activated phosphors. *The Journal of Chemical Physics* **48**:217–222.

Butler, K. H. 1947. Fluorescence of silicate phosphors. *Journal of the Optical Society of America* **37**:566–571.

Curie, D. 1960. *Luminescence in Crystals.* New York: John Wiley & Sons.

Dexter, D. L. 1952. A theory of sensitized luminescence in solids. *The Journal of Chemical Physics* **21**:836–850.

Duboc, C. A. 1956. Nonlinearity in photoconducting phosphors. *British Journal of Applied Physics, Supplement* **4**:107–111.

Engle, D. G. and Hopkins, B. S. 1925. Studies in luminescence. *Journal of the Optical Society of America and the Review of Scientific Instruments* **11**:599–606.

Etzel, H. W., Schulman, J. H., Ginther, R. J., and Claffy, E. W. 1952. Silver-activated alkali halides. A letter in *Physical Review* 1063–1064.

Fonda, G. R. 1939. Characteristics of silicate phosphors. *Journal of Physical Chemistry* **43**:561–577.

Fonda, G. R. 1940. The preparation of fluorescent calcite. *Journal of Physical Chemistry* **44**:435–439.

Fonda, G. R. 1940. The yellow and red zinc silicate phosphors. *Journal of Physical Chemistry* **44**:851–861.

Fonda, G. R. 1949. The enigma of multiple band emission. *Journal of the Electrochemical Society* **96**:4242–44.

Fonda, G. R. 1950. Dependence of emission spectra of phosphors upon activator concentration and temperature. *Journal of the Optical Society of America* **40**:347–352.

Fonda, G. R. 1956. Energy transfers in the calcium halophosphate phosphors. *British Journal of Applied Physics, Supplement* **4**:69–73.

Fonda, G. R. 1956. Two arsenate phosphors and the significance of their emission. *Journal of the Electrochemical Society* **103**:400–403.

Froelich, H. C. 1948. Manganese activated calcium silicate phosphors. *Journal of the Electrochemical Society* **33**:101–113.

Gallivan, J. B. and Deb, S. K. 1973. Photoluminescence of mercurous halides. *Journal of Luminescence* **6**:77–82.

Garlick, G. F. J. and Gibson, A. F. 1948. The electron trap of luminescence in sulphide and silicate phosphors. *Physical Society Proceedings* **60**:574–590.

Garlick, G. F. J. 1949. *Luminescent Materials.* Oxford: Clarendon Press.

Garlick, G. F. J. 1956. Absorption, emission and storage of energy in phosphors. *British Journal of Applied Physics, Supplement* **4**:85–90.

Gobrecht, H. and Weiss, W. 1955. Lumineszenzuntersuchungen an Uran-aktivierten Erdalkaliwolframaten und -molybdaten. *Zeitschrift fur Physik* **140**:139–149.

Goldberg, P. 1966. *Luminescence of Inorganic Solids.* New York: Academic Press.

Goldschmidt, V. M. 1954. *Geochemistry.* Oxford: Clarendon Press.

Gorbenko-Germanov, D. S. and Zenkova, R. A. 1964. On the vibrational structure of the ground and excited levels of UO_2^{++} in K_4 $[UO_2(CO_3)_3]$. *Optics and Spectroscopy* **20**:467–469.

Groenink, J. A. and Blasse, G. 1979. Some new observations on the luminescence of $PbMoO_4$ and $PbWO_4$. *Journal of Solid State Chemistry* **32**:9–20.

Haberlandt, H., Hernegger, F., and Scheminzky, F. 1949. Die Fluoreszenzspektren von Uranmineralien im filtrierten ultravioletten Licht. *Spectrochimica Acta* **4**:21–35.

Halsted, R. E., Apple, E. F., and Prener, J. S. 1959. Two-stage optical excitation in sulfide phosphors. *Physical Review Letters* **2**:420–421.

Hensler, J. R. 1959. Chemistry: synthesis of colour centres in silica and their thermoluminescence. Letter in *Nature* **183**:672–673.

Hummel, F. A. 1961. Cordierite-indialite: a new manganese-activated phosphor. *Journal of the Electrochemical Society* **108**:809–810.

Hunt, B. E. and McKeag, A. H. 1959. Copper and tin-activated halophosphate phosphors. *Journal of the Electrochemical Society* **106**:1032–1036.

Jaffe, P. M. 1964. Iron activated ZnS phosphors. *Electrochemical Society* **111**: 52–61.

Jenkins, H. G., McKeag, A. H., and Ranby, P. W. 1949. Alkaline earth halophosphates and related phosphors. *Electrochemical Society Journal* **96**:1–12.

Klasens, H. A. 1953. On the nature of fluorescent centers and traps in zinc sulfide. *Electrochemical Society Journal* **100**:72–80.

Klasens, H. A., Hoekstra, A. H., and Cox, A. P. M. 1957. Ultraviolet fluorescence of some ternary silicates activated with lead. *Electrochemical Society Journal* **104**:93–100.

Klick, C. C. 1957. Divalent manganese as a luminescent centre. *British Journal of Applied Physics, Supplement* **4**:74–78.

Klick, C. C. and Schulman, J. H. 1952. On the luminescence of divalent manganese in solids. *Journal of the Optical Society of America* **42**:910–916.

Koda, T. and Shionoya, S. 1964. Nature of the self-activated blue luminescence center in cubic ZnS:Cl single crystals. *Physical Review* **136**:541–555.

Kotera, Y., Yonemura, M., and Sekine, T. 1961. Activation by anions in the oxy-acid phosphors. *Journal of the Electrochemical Society* **108**:540–545.

Kreidl, N. J. 1945. Recent studies on the fluorescence of glass. *Journal of the Optical Society of America* **35**:249–257.

Kroger, F. A. and Bakker, J. 1941. Luminescence of cerium compounds. *Physica* **8**:628–646.

Kroger, F. A. 1947. Tetravalent manganese as an activator in luminescence. *Nature* **159**:706–707.

Kroger, F. A. 1947. Luminescence of solid solutions of the system $CaMoO_4$–$PbMoO_4$ and of some other systems. *Phillips Research Report* **2**:183–189.

Kroger, F. A. and Hellingman, J. E. 1948. The blue luminescence of zinc sulfide. *Journal of the Electrochemical Society* **93**:156–171.

Kroger, F. A. and Hoogenstraten, W. 1948. Decay and quenching of fluorescence in willemite. *Physica* **14**:425–441.

Kroger, F. A., Overbeek, J. T. G., Goorissen, J., and Boomgaard, J. van den. 1949. Bismuth as activator in fluorescent solids. *Electrochemical Society Journal* **96**:132–141.

Kroger, F. A. and Hellingman, J. E. 1949. Chemical proof of the presence of chlorine in blue fluorescent zinc sulfide. *Journal of the Electrochemical Society* **95**:68–69.

Kroger, F. A. 1949. A proof of the associated-pair theory for sensitized luminophors. *Physica* **15**:801–806.

Kroger, F. A. 1949. Sodium and lithium as activators of fluorescence in zinc sulfide. *Journal of the Optical Society of America* **39**:670–672.

Kroger, F. A. and Hoogenstraaten, W. 1949. Temperature quenching and decay of fluorescence in zinc-beryllium silicates activated with manganese. *Physica* **15**:557–568.

Kroger, F. A. and Hoogenstraaten, W. 1950. The location of dissipative transitions in luminescent systems. *Physica* **16**:30–32.

Kroger, F. A. 1949. *Some Aspects of the Luminescence of Solids.* New York: Elsevier.

Kroger, F. A. and Vink, H. J. 1953. The origin of the fluorescence in self-activated ZnS, CdS, and ZnO. *The Journal of Chemical Physics* 22:250–252.

Leverenz, H. W. 1944. Phosphors versus the periodic system of the elements. *Proceedings of the I.R.E.* 256–263.

Leverenz, H. W. 1968. *An Introduction to the Luminescence of Solids.* New York: Dover Publications Inc.

Lewis, G. N., Lipkin, D., and Magel, T. T. 1941. Reversible photochemical processes in rigid media: a study of the phosphorescent state. *Journal of the American Chemical Society* **63**:3005–3018.

Linwood, S. H. and Weyl, W. A. 1942. The fluorescence of manganese in glasses and crystals. *Journal of the Optical Society of America* **32**:443–453.

Makai, E. 1949. High valent manganese as activator of luminescence. *Journal of the Electrochemical Society* **95**:107–111.

Medlin, W. L. 1963. Emission centers in thermoluminescent calcite, dolomite, magnesite, aragonite, and anhydrite. *Journal of the Optical Society of America* **53**:1276–1285.

Meixner, H. von. 1940. Fluoreszenzanalytische, optische und chemische Beobachtungen an Uranmineralen. *Chem. Erde* **12**:433–450.

Merrill, J. B. and Schulman, J. H. 1948. The $CaSiO_3$:(Pb + Mn) phospor. *Journal of the Optical Society of America* **38**:471–479.

Millson, H. E. and Millson, E. M., Jr. 1950. Observations on exceptional duration of mineral phosphorescence. *Journal of the Optical Society of America* **40**(7):430–435.

Murata, K. J. and Smith, R. L. 1946. Manganese and lead as coactivators of red fluorescence in halite. *American Mineralogist* **31**:527–538.

Nichols, E. L. and Howes, H. L. 1926. Note of the rare earths as activators of luminescence. *Journal of the Optical Society of America and Review of Scientific Instruments* **13**:573–587.

Orgel, L. E. 1958. Phosphorescence of solids containing the manganous or Ferric ions. *Journal of Chemical Physics* **23**:1958.

Prener, J. S. and Williams, F. E. 1956. Activator systems in zinc sulfide phosphors. *Electrochemical Society Journal* **103**:342–346.

Pringsheim, P. and Vogel, M. 1943. *Luminescence of Solids and Liquids.* New York: Interscience Publishers.

Przibram, K. 1956. *Irradiation Colors and Luminescence.* London: Pergamon Press.

Przibram, K. 1949. The light emitted by europium compounds. Letter in *Nature* **163**:989.

Randall, J. T. and Wilkins, M. H. F. 1941. Phosphorescence and electron traps: I. The study of trap distributions. *Proceedings of the Royal Society of London* **184**:366–407.

Rankama, K. and Sahama, T. G. 1950. *Geochemistry.* Chicago: University of Chicago Press.

Schein, M. and Katz, M. L. 1936. Ultra-violet luminescence of sodium chloride. Letter in *Nature* **138**:883.

Schulman, J. H. 1946. Luminescence of $(Zn, Be)_2 SiO_4$:Mn and other manganese-activated phosphors. *Journal of Applied Physics* **17**:902–908.

Schulman, J. H., Evans, L. W., Ginther, R. J., and Murata, K. J. 1947. The sensitized luminescence of manganese-activated calcite. *Journal of Applied Physics* **18**:732–739.

Schulman, J. H., Ginther, R. J., and Klick, C. C. 1950. A study of the mechanism of sensitized luminescence of solids. *Journal of the Electrochemical Society* **97**:123–132.

Schulman, J. H. 1955. Physical measurements and the nature of the luminescent centers. *British Journal of Applied Physics, Supplement* **4**:64–69.

Seitz, F. 1938. Interpretation of the properties of alkali halide–thallium phosphors. *Journal of Chemical Physics* **6**:150–162.

Shionoya, S. 1955. Sensitized luminescence of zinc sulfide phosphors activated with copper and manganese. Letter in *Journal of Chemical Physics* **23**:1173.

Shionoya, S. 1955. Thermoluminescence of zinc sulfide phosphors doubly activated with copper and manganese. *Journal of Chemical Physics* **23**:1976–1977.

Studer, F. J. and Fonda, G. R. 1949. Optical properties of calcium silicate phosphors. *Journal of the Optical Society of America* **39**:655–660.

Studer, F. J. and Rosenbaum, A. 1949. The phosphorescence decay of halophosphates and other doubly activated phosphors. *Journal of the Optical Society of America* **39**:685–689.

Suzuki, A. and Shionoya, S. 1970. Evidence for the pair emission mechanism of the green-Cu luminescence in ZnS. Letter in *Journal of Luminescence* **3**:74–76.

Tanaka, T. 1924. On the cathodo luminescence of solid solutions of forty-two metals. *Optical Society of America Journal* **8**:287–318.

Williams, F. E. 1949. Review of the interpretations of luminescence phenomena. *Journal of the Optical Society of America* **39**:648–654.

Williams, F. E. 1955. Theory of activator systems in luminescent solids. *British Journal of Applied Physics* **6**:97–102.

Williams, F. E. and Eyring, H. 1955. The mechanism of the luminescence of solids. *The Journal of Chemical Physics* **15**:289–304.

CHAPTER 9

Bouma, P. J. 1947. *The Physical Aspects of Color.* Eindhoven, Netherlands: N. V. Philips Gloeilampenfabrieken.

Committee on Colorimetry, 1953. *The Science of Color.* Washington, D.C.: Optical Society of America.

Kelly, K. L. and Judd, D. B. 1976. *Color – universal language and dictionary of names. National Bureau of Standards (U.S.) Special Publication 440.*

Le Grand, Y. 1962. *Light, Color and Vision.* London: Chapman and Hall Ltd.

Munsell Color Company 1942. *Munsell Book of Color.* Baltimore, Md.: Munsell Color Co.

Murray, H. D. (Ed.). 1952. *Color in Theory and Practice.* London: Chapman and Hall Ltd.

Newsome, D. and Modreski, P. 1981. The colors and spectral distributions of fluorescent minerals. *Journal of the Fluorescent Mineral Society.* **10**:1–7.

Ridgway, R. 1912. *Color Standards and Color Nomenclature.* Washington, D.C.: published by author.

Wright, W. D. 1958. *The Measurement of Color.* New York: Macmillan.

Index